RAND

T0159547

The Environmental Implications of Population Dynamics

Lori M. Hunter

Supported by the
William and Flora Hewlett Foundation
David and Lucile Packard Foundation
Rockefeller Foundation

A RAND Program of Policy-Relevant Research Communication

The research described in this report was supported by the David and Lucile Packard Foundation, the William and Flora Hewlett Foundation, and the Rockefeller Foundation.

Library of Congress Cataloging-in-Publication Data

Hunter, Lori M.
    The environmental implications of population dynamics / Lori M. Hunter.
    p. cm.
    "MR-1191-WFHF/RF/DLPF."
    Includes bibliographical references.
    ISBN 0-8330-2901-0
    1. Population—Environmental aspects. I. Labor and Population Program. II. Population Matters (Project) III. Title.

HB849.415 .H863 2000
337.7—dc21

                                       00-062769

Published 2000 by RAND
1700 Main Street, P.O. Box 2138, Santa Monica, CA 90407-2138
1200 South Hayes Street, Arlington, VA 22202-5050
RAND URL: http://www.rand.org/
To order RAND documents or to obtain additional information, contact Distribution Services: Telephone: (310) 451-7002;
Fax: (310) 451-6915; Internet: order@rand.org

PREFACE

It has become increasingly clear that human populations have a powerful effect on the environment. Yet the exact relationship between population dynamics and the environment is complex and not well understood. This report draws from demographic and environmental literature to examine what is known about the association between population dynamics and the natural environment.

The work was conducted in RAND's Labor and Population program as part of the *Population Matters* project. The primary focus of *Population Matters* is synthesizing and communicating the findings and implications of existing research in ways that policy analysts and others will find accessible.

The *Population Matters* project is funded by grants from the William and Flora Hewlett Foundation, the David and Lucile Packard Foundation, and the Rockefeller Foundation. This document should be of interest to anyone concerned with demographic issues, environmental issues, and the interrelationships among them. For a list of our publications, please consult the inside back cover of this report. For further information on the *Population Matters* project, contact:

Julie DaVanzo, Director, *Population Matters*
RAND
P.O. Box 2138
1700 Main St.
Santa Monica, CA 90407-2138
Julie_DaVanzo@rand.org

Or visit the project's website at http://www.rand.org/popmatters.

CONTENTS

FIGURES

TABLES

The Earth's population doubled between 1960 and 1999, increasing from three billion to six billion people. During that period, human-induced changes in the global environment accelerated in unprecedented fashion. Given continued population growth and environmental degradation, it has become paramount that we deepen our understanding of the role played by human population dynamics in environmental change. Drawing from the scientific literature, this report presents a synthesis of what is known about the role played by human population factors in environmental change. Specifically, the report discusses the following:

- The relationship between population factors—size, distribution, and composition—and environmental change.

- The primary forces that mediate this relationship: technology, the institutional and policy contexts, and cultural factors.

- Two specific aspects of environmental change that are affected by population dynamics: climate change and land-use change.

- Implications for policy and further research.

ENVIRONMENTAL IMPLICATIONS OF POPULATION SIZE, DISTRIBUTION, AND COMPOSITION

Population Size

Population size is inherently linked to the environment as a result of individual resource needs as well as individual contributions to pollution. However, no simple relationship exists between population size and environmental change. Sheer human numbers in some instances have a direct impact on the environment. More often, however, the environmental implications of population size are ultimately determined by complex interactions among many forces, including technology, political and institutional contexts, and cultural factors.

However, as global population continues to grow, limits on such global resources as land and water have come into sharper focus. For example, only in the latter half of the twentieth century has the unavailability of land become a potentially limiting factor in global food production. Assuming constant rates of production, per capita land requirements for food production now fall within the range of estimated available cultivable land. Likewise, continued population growth occurs in the context of an accelerating human thirst for water: Global water consumption rose sixfold between 1900 and 1995, more than double the rate of population growth.

Population size also influences pollution levels in complex ways. Though again this interaction is difficult to gauge, researchers have tried to calculate the relationship between population growth and pollution increases. Studies of air pollution in California, for instance, suggest that a 10 percent increase in population at the county level produces an emissions increase of 7.5 to 8 percent for pollutants associated with automobile exhaust, largely because local population growth is important as a determinant of the volume of consumption. Greater numbers of people, for instance, typically imply more cars.

Population Distribution

"Population distribution" refers to the dispersal and density of population. During the past 40 years, two trends have powerfully influenced the distribution of humans around the globe. First, continued high fertility rates in many developing regions, coupled with low fertility in more-developed regions, have resulted in ever-increasing shares of the global population residing in less-developed countries. According to UN estimates, 80 percent of the world population in 1999 lived in developing nations. Second, the Earth's population is increasingly concentrated in urban areas. As recently as 1960, only one-third of the world's population lived in cities. By 1999, the percentage had increased to nearly half (47 percent). This trend is expected to continue well into the twenty-first century.

The distribution of people around the globe has three main implications for environmental change. First, as less-developed regions cope with an increasing share of global population, pressures will intensify on already dwindling resources within many of these areas. Second, the redistribution of population through migration shifts the relative pressures exerted on local environments, perhaps easing the strain in some areas and increasing it in others. Finally, the trend toward urbanization poses particularly complex environmental challenges. The rapid pace of urbanization often hinders the development of adequate infrastructure and regulatory mechanisms for coping with pollution and other byproducts of growth, often resulting in high levels of air and water pollution and other environmental ills. Furthermore, urbanization can alter local climate patterns. Concentrations of artificial surfaces, such as brick and concrete, can create "heat islands." In cities with more than 10 million people, the mean annual minimum temperature may be as much as 4 degrees Fahrenheit higher than in nearby rural areas. In addition, poorly planned urban development—"sprawl"—can result in loss of agricultural land and natural habitat.

Population Composition

"Population composition" refers to the characteristics of a particular group of people. Age and socioeconomic composition, for instance, have environmental implications.

As for age composition, owing to the population boom of recent decades and increased longevity across the globe, today's human population has both the largest cohort of young people (age 24 and under) and the largest proportion of elderly in history. Understanding population characteristics helps illuminate some of the mechanisms through which population dynamics affect environmental conditions. For example, migration propensities vary by age. Young people are more likely than their older counterparts to migrate, primarily as they leave the parental home in search of new opportunities. Given the relatively large younger generation, we might anticipate increasing levels of migration and urbanization and, therefore, intensified urban environmental concerns.

Income is an especially important demographic characteristic relevant to environmental conditions. Across nations, the relationship between economic development and environmental pressure resembles an inverted U-shaped curve; nations with economies in the middle-development range are most likely to exert powerful pressures on the natural environment, mostly in the form of industrial emissions. By contrast, the least-developed nations—because of low levels of industrial activity—are likely to exert relatively lower levels of environmental pressure. In addition, at highly advanced development stages, environmental pressures should subside due to improved efficiencies.

Within countries and across households, the relationship between income and environmental pressure is different. Environmental pressures can be greatest at the lowest and highest income levels. Population growth and poverty often interact to produce unsustainable levels of resource use. Furthermore, higher levels of income tend to correlate with increased levels of production and consumption.

MEDIATING FACTORS: TECHNOLOGY, INSTITUTIONAL AND POLICY CONTEXTS, AND CULTURAL FACTORS

Several factors mediate the relationship between human population dynamics and the natural environment. Aspects of society relating to current technology, institutions, policy, and culture alter the ways in which demographic and environmental factors interact.

Technology

Technological factors have always influenced the relationship between population dynamics and environmental change. In some cases, technological advancements have caused greater environmental change than demographic trends alone would have led us to expect. The agricultural revolution of the seventeenth and eighteenth centuries, for instance, enabled demographic shifts that

otherwise could not have occurred because it permitted food production sufficient to feed the world's growing population.

The technological changes that have most affected environmental
conditions relate to energy use. In particular, the consumption of oil,
natural gas, and coal increased dramatically during the twentieth
century. Until about 1960, developed nations were responsible for
most of this consumption. Since then, however, the newly developing nations have experienced increasing levels of industrialization,
resulting in greater reliance upon resource-intensive and highly
polluting production processes. Obviously, improved energy efficiencies could greatly diminish the environmental impacts from
energy consumption in both developed and developing nations.

Institutional and Policy Contexts

Institutions and policy responses are significant mechanisms
through which humans react to environmental change and, in so
doing, affect subsequent environmental change. These mechanisms
can operate for good or ill. For example, following the Montreal
Protocol of 1987, limits were established for emissions of chlorofluorocarbons (CFCs), which cause ozone depletion. The ozone layer
shields humans from the sun's high-energy ultraviolet radiation. As
a result of the emission policy, CFC consumption has fallen by nearly
70 percent, and the ozone layer is expected to return to normal by
the middle of this century. While demographic factors were not the
only environmentally destructive forces in this example, population
size influenced environmental conditions by being a market for CFC-
producing goods. People buy refrigerators. In this instance, however, it was the policy response that ultimately defined the relationships among technology, consumption, population, and
environmental change.

Although policy actions can ameliorate environmental decline, on
occasion misguided policy may become a powerful force behind
degradation. The ecological and social dilemmas facing the Aral Sea
basin offer one extreme example of the effects of policies regarding
resource use. The sea basin in central Asia is shared by several
nations, mainly Uzbekistan and Kazakhstan. Since 1960, the Aral Sea
has shrunk forty percent and has become increasingly contaminated.
Although some of the decline and contamination stems from natural

variations, research has demonstrated that human forces have been the primary cause behind the ecological destruction—in particular, irrigation policies of the former Soviet Union appear to blame. In this case, the role of local population in environmental decline was shaped by policies regarding water use.

Cultural Factors

Cultural factors can also play a role in how population dynamics affect the environment. As examples, cultural differences with respect to consumption patterns and attitudes toward wildlife and conservation are likely to affect how populations interact with the environment. For instance, one study demonstrated distinctive patterns with regard to attitudes, knowledge, and behavior toward wildlife across three industrial democracies. While Americans and Germans express broad appreciation for a variety of animals, Japanese culture emphasizes the experience of nature in controlled, confined, and highly idealized circumstances (e.g., bonsai, rock gardening, flower arranging). These cultural variations in turn influence conservation strategies, because public support for various policy interventions will reflect societal values.

TWO SPECIFIC ARENAS OF POPULATION-ENVIRONMENT INTERACTION: GLOBAL CLIMATE CHANGE AND LAND-USE PATTERNS

Two specific areas of inquiry help to illustrate the challenges of understanding the complex influence of population dynamics on the environment: global climate change and land-use patterns. On the other hand, these examples also demonstrate the growing body of scientific evidence that illuminates the interrelationships between demographics and environmental context.

Global Climate Change

Recent years have been among the warmest on record. Evidence suggests that temperatures have been influenced by growing concentrations of greenhouse gases, such as carbon dioxide, which absorb solar radiation and warm the Earth's atmosphere. To what extent can climate change be attributed directly to demographic factors? A growing body of evidence suggests that many of the changes in atmospheric gas are human-induced. The demographic influence appears primarily in three forms: contributions to CO_2 emissions stem from fossil fuel use related to industrial production and energy consumption; land-use changes, such as deforestation, also affect the exchange of carbon dioxide between the Earth and the atmosphere; and other consumption-related processes, such as rice paddy cultivation and livestock production, are responsible for greenhouse-gas releases to the atmosphere, particularly methane.

Research has shown that population size and growth are important factors in the emission of greenhouse gases. One study concludes that population size and growth will account for 35 percent of the global increase in CO_2 emissions between 1985 and 2100 and 48 percent of the increase from developing nations during that period. However, as population growth slows during the next century, its contribution to emissions is expected to decline. This decline will be especially large in the context of developing nations. While population-driven emissions from developed nations are estimated to contribute 42 percent of CO_2 emissions between 1985 and 2020, they are expected to contribute only 3 percent between 2025 and 2100.

Land Use

Fulfilling the resource requirements of a growing population ultimately requires some form of land-use change, whether to expand food production through forest clearing, to intensify production on already cultivated land, or to develop the infrastructure needed to support increased population. Indeed, it is humans' ability to manipulate the landscape that has allowed for the rapid pace of contemporary population growth.

Agriculture and deforestation are two prominent forms of human-induced land-use change. During the past three centuries, the amount of Earth's cultivated land has grown by more than 450 percent, increasing from 2.65 million square kilometers to 15 million square kilometers. At the same time, the world's forests have been shrinking. Deforestation is closely linked to agricultural land-use change, because it often represents a consequence of agricultural expansion. A net decline in forest cover of 180 million acres occurred during the 15-year interval 1980–1995, although changes in forest cover vary greatly across regions.

Changing land use and deforestation in particular have several ecological impacts. Agriculture can lead to soil erosion, while overuse of chemical inputs can also degrade soil. Deforestation also increases soil erosion, in addition to reducing rainfall due to localized climate changes, lessening the ability of soils to hold water, and increasing the frequency and severity of floods. Land-use change in general results in habitat loss and fragmentation—the primary cause of contemporary species decline. It has been suggested that if current rates of forest clearing continue, a quarter of all species on Earth could be lost within the next 50 years.

IMPLICATIONS FOR POLICY

The environmental implications of demographic dynamics are complicated and can sometimes be controversial. While some view population growth in developing regions as the primary culprit in environmental decline, others focus on the costly environmental effects of overconsumption among the developed nations. Such differing emphases can lead to a disagreement over the most effective and equitable policy solutions—slow population increase in less-developed nations or lessen destructive production and consumption patterns of the more-developed nations?

Such a debate, however, presumes that a one-step solution to the complex realities of the relationship between population and the environment exists. Both population growth and consumption play a role in environmental change and are among the many factors that should be considered and incorporated into realistic policy debate and prescriptions. Other demographic factors, such as population distribution and composition, are also relevant.

Some specific implications for policy that emerge from reviewing the scientific literature include the following:

- Family planning policies that enable couples to avoid unwanted pregnancies would reduce fertility and rates of population growth, therefore reducing pressure on environmental resources.

This would be particularly beneficial in areas already characterized by resource scarcity.

- Rural development policies could reduce rural-to-urban migration, perhaps easing pressure on urban infrastructure. This would particularly benefit areas where rural resource shortages or lack of opportunities fuel rapid urban growth.

- More equitable land-tenure policies could ease resource pressures and, therefore, reduce agricultural expansion and rural-to-urban migration. This would especially benefit areas with subsistence agriculture where individuals lack access to land.

- Policies encouraging sustainable intensification of land resources could increase yields, thereby lessening the need for agricultural extensification. This would be especially beneficial in areas characterized by arable land shortages.

- Policies providing incentives for the development of sustainable production processes could ease environmental pressure. Both developed and developing regions could benefit from the application of improved technological efficiencies.

- Policies providing education and encouragement for sustainable consumption could ease environmental pressure. This is particularly true in areas where consumption and production processes are environmentally intensive.

- Careful planning must accompany change in local population densities. This is true both in rapidly growing megacities as well as in less-densely populated areas receiving large influxes of migrants.

Population policies represent only one of the many possible responses to the environmental implications of demographic dynamics. Yet population does matter, and increased attention to the associations between the environment and population size, distribution, and composition can improve policy capacity to respond to contemporary environmental change.

RESEARCH NEEDS

A more precise scientific understanding of the complex interaction between demographic processes and the environment is needed. To accomplish this goal, natural and social scientists must work together. This integration will continue the encouraging movement toward a truly interdisciplinary global environmental science. The use of such modern technology as remote sensing to study environmental change across time is especially promising.

Scientists must also continue to collect data and develop new models that link natural and social processes. Research on population and the environment has continually been hampered by a lack of appropriate data, and concerns persist with regard to data availability and quality. Support must be provided for the collection and provision of relevant information allowing development of local, regional, national, and global scale models of the interrelations between natural and social processes.

Finally, researchers can work to inform relevant policy circles. For instance, micro-level studies of local area interactions between population and the environment provide important insight into local processes and relevant policy. However, macro-scale analyses of global processes provide critical information for use within the international policy context.

ACKNOWLEDGMENTS

Julie DaVanzo, Director of RAND's *Population Matters* project, has been especially central to the development of this report, providing insightful guidance on both general structure and specific content. Her skill at objective review of scientific literature is admirable. David Adamson, RAND Communications Analyst, drafted the report's summary and has provided important input from start to finish. Early reviews were provided by Mark Bernstein, Sara Curran, Ezekial Kalipeni, and Anne Pebley. John Bongaarts and Ron Rindfuss provided especially important comments on later versions. The author also thanks colleagues at Utah State University for their support during the writing of this report. And finally, I am indebted to Julie DaVanzo and Anne Pebley for recognizing the centrality of the intersection of demographic and environmental issues to human well-being, providing justification for this review as a component of the *Population Matters* series. Thank you for the opportunity to contribute.

CFCs	Chlorofluorocarbons
FAO	Food and Agricultural Organization
IPCC	International Panel on Climate Change
MLC	Mono Lake Committee
NGO	Nongovernmental organization
Ppmv	Parts per million by volume
PRB	Population Reference Bureau
PRC	People's Republic of China
UN	United Nations
UNEP	UN Environment Programme
UNFPA	UN Population Fund
USDA	U.S. Department of Agriculture
WCED	World Commission on Environment and Development
WCMC	World Conservation Monitoring Center
WHO	World Health Organization
WMO	World Meteorological Organization
WRI	World Resources Institute

Chapter One

INTRODUCTION

Rapid population growth and global environmental change are two topics that have received substantial public attention over the past several decades. Population increase became a global public policy issue during the mid-twentieth century as mortality declines in many developing nations were not matched with reductions in fertility, resulting in unprecedented growth rates. Concern with environmental change has come to the forefront primarily since 1970, with discernible levels of environmental degradation fueling public concern with the scope of contemporary environmental transformations and the advent of satellite imagery aiding environmental research (Clarke, 1996).

Today, both population and environmental concerns are often subsumed within the dialogue on "sustainable development," aimed at addressing the needs and aspirations of today's population without compromising the well-being of future generations (WCED, 1988). Responsible stewardship of the Earth's natural resources is inherent in the idea of sustainability. One approach to lessening environmental pressure may be to diminish the impact of humans through population stabilization—the assumption being that fewer people means less pressure on land, air, and water environments. But is it so simple?

The following report reviews the environmental implications of population dynamics, based on current knowledge of the relationships between population factors and various aspects of the natural environment. This inquiry is important for several reasons:

- **Compared with human-induced environmental changes of centuries past, the geographic scope of contemporary human-induced change is much larger. The rate of change is also much faster.** Climate change provides an example. The Earth's surface air temperature has increased by between 0.3 degrees and 0.6 degrees Celsius since the late nineteenth century, and recent years have been among the warmest since the beginning of instrumental recordkeeping around 1860 (IPCC, 1995a). In seeking answers to this global warming, scientists have discovered that such human activities as land-use change and fossil fuel use appear linked to increased concentrations of several atmospheric gases which warm the Earth's atmosphere and surface temperature (IPCC, 1995a). As a result, humans appear partially responsible for environmental changes at the global scale—large-scale changes that have taken place in less than 200 years.

- **Some human-induced environmental changes have irreversible consequences.** Consider species extinction: it is estimated that nearly 12 percent of mammals and 11 percent of bird species are currently threatened with extinction (WCMC, 1992). In fact, the currently confronted episode of biodiversity loss is greater than the world has experienced for the past 65 million years (Raven and McNeely, 1998). Although some human activities, such as conservation programs, aim to preserve or enhance biodiversity, human-induced habitat destruction remains the primary cause of species decline (Southwick, 1996; Wilson, 1992).

- **The effects of contemporary environmental changes on humans are also increasing in scope.** In other words, not only does population affect the environment, but also the reverse is true. For example, global estimates suggest there may be as many as 25 million "environmental refugees"—individuals who have migrated because they can no longer secure a livelihood from the land because of deforestation, desertification, soil erosion, and other environmental problems. The most significant environmental migrations have occurred in Sub-Saharan Africa, the Indian subcontinent, China, Mexico, and Central America (Myers, 1997).

- **Human population continues to grow.** In the face of rapid, large-scale, human-induced environmental change, human numbers increase each year by approximately 80 million (United

Nations, 1998a). Unfortunately, those areas with continuing high fertility and resulting population growth are typically those with the least environmental resilience and those facing the greatest resource constraints. Although many developing nations already face severe air pollution, water contamination, and other environmental ills, more than 90 percent of future population growth is projected to take place in these areas.

All told, the scale, scope, and irreversibility of contemporary environmental change and the role of population in that change attest to the relevance of population factors to the environment. Yet questions remain about the specific nature of the relationship between population and the natural environment. For instance:

- Is population size the most important demographic concern with regard to environmental impacts? How important are other demographic factors (e.g., population distribution, composition)?

- Do the environmental implications of demographic factors vary across contexts? Across time?

- What are the nondemographic factors that play roles in determining population's environmental influence? How important is population per se relative to these other factors?

- Does population appear especially influential with regard to any specific aspects of the environment?

- What policy responses are appropriate given our present understanding of the relationship between population and the environment?

In this report, we review demographic trends and scientific research regarding the environmental implications of population dynamics to address these questions.

A FRAMEWORK FOR CONSIDERING THE RELATIONSHIP BETWEEN POPULATION AND THE ENVIRONMENT

Discussing the relationship between population and the environment is not simple. "Population" is a multidimensional concept that can relate to the size, distribution, density, or composition of an area's inhabitants. "Environment" is no less complex—encompassing qualities of the air, water, and land on which humans and all other species depend. Further complicating the relationship between population and the environment are the myriad "mediating" influences that ultimately shape this association. These include technological factors (e.g., forms of energy production), political factors (e.g., environmental regulation), and cultural factors (e.g., attitudes toward wildlife and conservation).

Figure 1.1 presents a conceptual framework describing the relationship between population and the environment in fairly simple terms. The framework helps organize the following review of this complex relationship; each of the following chapters relates to a component. Demographic factors are covered in Chapter Two (population size and growth), Chapter Three (population distribution), and Chapter Four (population composition). In each case, current trends are reviewed and their environmental implications are discussed. Chapter Five presents an overview of several "mediating factors" that influence the relationship between population and the environment. Chapter Six relates to the right-hand side of the framework, where

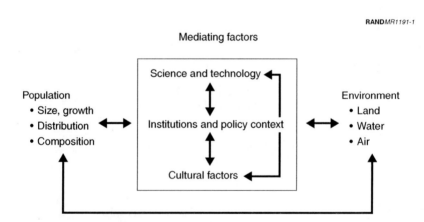

RAND*MR1191-1*

SOURCE: Adapted from MacKellar et al., 1998.

Figure 1.1—Framework for Considering the Relationship Between Population and the Environment

the environmental implications are represented in this report by two select topics: climate change and land-use change. Finally, Chapter Seven offers a review of policy implications and a discussion of research necessary to fill the gaps in our knowledge of the relationship between population and the environment. An especially important lesson from this review is that the complexity of the relationship between population and the environment necessitates many different types of policy response—population-oriented policies represent only one of many routes through which modern societies must respond to human-induced environmental decline.

This report is not intended to provide a comprehensive review of existing knowledge on the relationships between population and the environment, because that is simply beyond the scope of this project. Instead, we aim to provide a review of the evidence best allowing determination of the environmental implications of population dynamics. To this end, the focus here is on the influence of demographic factors on the environment, as opposed to the reciprocal aspect of this relationship, which would encompass the demographic implications of environmental change. Consider, for instance, the relevance of environmental pollution to human mortality, or the potential relationship between resource shortages and out-migration. While these relationships are indeed interesting and important, they require a manuscript of their own. References for readings along these lines are provided in the appendix.

Chapter Two

POPULATION SIZE: TRENDS AND ENVIRONMENTAL IMPLICATIONS

Population size is inherently linked to the environment as a result of individual resource needs as well as individual contributions to pollution. As a result, population growth yields heightened demands on air, water, and land environments, because they provide necessary resources and act as sinks for environmental pollutants.

Population policies designed to reduce future growth represent logical responses to the environmental implications of population size, although fertility reduction cannot be seen as sufficient response to contemporary human-induced environmental change. A decrease in human numbers does not necessarily suggest a decrease in environmentally significant behaviors. Furthermore, the assumption that each additional individual has an equal impact on resources is too simplistic. Factors related to both the individual and to the social and environmental contexts will determine the ultimate nature of the relationship.

For example, the cultural context into which an individual is born will influence that individual's relationship with the environment—empirical evidence suggests that a child born in the United States will produce 10 times the pollution of a child born in Bangladesh (Stern et al., 1997). Much of this is the product of consumption patterns, where income-driven lifestyle changes increase the amount of energy and materials consumed.[1] One study suggests that, on aver-

---

[1] The environmental implications of income are further discussed in Chapter Four.

age, each American consumes more than 50 kilograms (approximately 110 pounds) of material per day, excluding water. The vast majority of this includes the materials required for the production and distribution of consumer goods (Wernick, 1997).

POPULATION SIZE TRENDS AND PATTERNS

In a demographic sense, the past three centuries have been unique compared to any other historical period. During prehistoric time, covering 99 percent of human history, population increase was very slow. However, within the past 350 years, human population has dramatically increased, with most estimates indicating that about 90 percent of the world's population growth has occurred during this period (Whitmore et al., 1990; see Figure 2.1). Advances in public health and hygiene brought about by the scientific-technological-economic revolution that began in the mid-eighteenth century particularly altered the course of global population history through

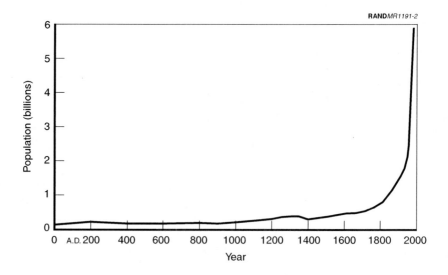

SOURCE: Adapted from Cohen, 1995, p. 82.

Figure 2.1—Estimated Human Population from 1 A.D. to the Present

dramatic reductions in mortality.[2] This led to staggering increases in the rate of population growth.

Today's world population continues to grow, although declines in fertility rates have made the rate of growth less than during the past two decades. According to recent United Nations estimates, global population currently increases by approximately 80 million—the size of Germany—each year and passed the 6 billion mark in late 1999. This is indeed a milestone, as 6 billion represents twice the global population in 1960 (United Nations, 1998a). The continuing growth is partly a result of population momentum gathered during the rapid growth of the 1980s and beyond, whereby the large number of individuals born during that decade are now entering their childbearing years, therefore practically ensuring continued growth at least through the middle of the next century.

Although fertility rates have declined in most areas, population growth continues to be fueled by high levels of fertility in many countries. In numerous African nations, the average number of children a woman would be expected to have, given current fertility levels, remains above 6.0—for example, 6.9 in Uganda, 7.0 in Somalia, and as high as 7.5 in Niger (PRB, 2000). Many Middle Eastern nations also face continued high fertility rates—for example, the average number of children expected per woman is 6.4 in Saudi Arabia, 6.5 in Yemen, and 7.1 in Oman (PRB, 2000). Variations in fertility levels reflect differences in the level of social and economic development, along with variations in contraceptive use. While an estimated 60 percent of married women in developed nations use modern contraceptive methods, only 10 percent do so in Yemen and only 8 percent in Uganda (PRB, 2000).

Declines in mortality have also played an important role in high levels of population growth. Low-cost technologies, such as antibiotics, vaccines, and improvements in nutrition and agriculture, have reduced mortality throughout the world. In less-developed regions, crude mortality rates decreased substantially from 24.2 deaths per 1,000 population in 1950–1955 to 9.1 in 1990–1995 (United Nations,

---

[2]A more detailed discussion of long-term demographic trends is provided by Demeny (1990). For additional information, Whitmore et al. (1990) discuss three interpretations of population trends over the last 3 million years.

1998c). There is a caveat, however, to these generally promising mortality trends. Recent estimates suggest a devastating toll from HIV/AIDS, particularly in several African nations. In the 29 hardest-hit African nations, for the period 2000–2005 life expectancy at birth (47.4 years) is projected to be seven years less than would be expected in the absence of AIDS (United Nations, 1999).[3] Even in the worst cases, however, overall population declines are not anticipated because of high fertility levels (United Nations, 1999)—the population of those 29 African countries is projected to increase by more than half during 1995–2015, from 446 million to 698 million (United Nations, 1999). This represents a doubling of population in the 30 year period from 1985–2015 (United Nations, 1999).[4]

The trends in fertility and mortality combine to yield the population projections presented in Figure 2.2. According to medium-fertility projections by the United Nations Population Division, world population could reach 8.9 billion in 2050 and may ultimately stabilize at nearly 11 billion around 2200 (United Nations, 1998a, 1998d).[5] This represents a near doubling of the current world population.

POPULATION SIZE AND THE ENVIRONMENT

Concern with the ecological effects of population size is certainly not new. British economist Thomas Malthus warned of the unsustain-

---

[3]Botswana has the highest prevalence of HIV, with nearly one of every four adults infected (22 percent of adult population). Life expectancy at birth there is projected to fall from 61 years in 1990–1995 to 47 years by 2000–2005 (United Nations, 1999).

[4]In *World Population Prospects, the 1998 Revision,* the United Nations estimated the demographic impact of HIV/AIDS for 34 countries that had at least 1 million population in 1995 and that had an estimated adult HIV seroprevalence of 2 percent or higher in 1997. Brazil and India are also included in the estimates of demographic impact because of their large populations, although their seroprevalence is not yet 2 percent. Twenty-nine of the countries considered are in Africa (Benin, Botswana, Burkina Faso, Burundi, Cameroon, Central African Republic, Chad, Congo, Cote d'Ivoire, Democratic Republic of the Congo, Eritrea, Ethiopia, Gabon, Guinea-Bissau, Kenya, Lesotho, Liberia, Malawi, Mozambique, Namibia, Nigeria, Rwanda, Sierra Leone, South Africa, Togo, Uganda, United Republic of Tanzania, Zambia, and Zimbabwe), three are in Asia (Cambodia, India, Thailand), and two are in Latin America and the Caribbean (Brazil, Haiti) (United Nations, 1999).

[5]Medium-fertility projections assume fertility will stabilize at replacement levels of slightly above two children per woman by 2050 (PRB 1998b; United Nations, 1998c).

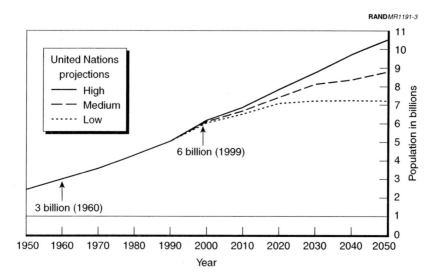

RAND*MR1191-3*

SOURCE: UNFPA, 1999.

**Figure 2.2—World Population Size According to Main Fertility
Scenarios, 1950–2050**

ability of unchecked population growth more than 200 years ago,
arguing that human population has a tendency to exceed the ability
of the environment to provide subsistence. In particular, Malthus
suggested that unrestrained population growth would outstrip the
ability of the Earth to provide sufficient food. Although this perspective has been criticized for its simplistic focus on population size as
the sole driving force in resource change,[6] Malthus initiated a debate
on carrying capacity and his influence continues today.[7] Many contemporary population-oriented interest groups focus on population
size as the determining factor in environmental degradation
(Campbell, 1998). In addition, a Neo-Malthusian outlook, focusing

---

[6]Others argue for consideration of technology (Boserup, 1965, 1976, 1981), consumption (Stern et al., 1997), and global market forces (Simon, 1981). Chapter Five provides
a general review of such mediating factors.

[7]Carrying capacity refers to the maximum population of a given organism that a particular environment can sustain. The term has its origin in the ecological sciences,
although it has been applied to human populations as well (e.g., Cohen, 1995).

on the role of population size and growth in environmental decline, appears in the classical economic and natural science perspectives in contemporary debates on the relationship between population and the environment (Jolly, 1994).

Global population size is inherently connected to land, air, and water environments because each and every individual uses environmental resources and contributes to environmental pollution. While the scale of resource use and the level of wastes produced vary across individuals and across cultural contexts, the fact remains that land, water, and air are necessary for human survival.

As for resource consumption, two commonsense points can highlight the implications of population size and growth. First, each person obviously requires food, the production of which typically requires land for agriculture or other forms of sustenance production.[8] Globally, about 1.5 billion hectares are cultivated for agriculture, representing the most suitable of an estimated 2 billion to 4 billion hectares characterized as cultivable (Southwick, 1996). To consider future land requirements in the face of increasing human population, Figure 2.3 presents the hectares required to meet the food demands of projected global population, assuming constant per capita production. Although there has been an excess of potentially cultivable land throughout human history, the exponential growth of human population has hastened the pace of land-use change. Hectares required for global food production now fall near the lower limit of estimated cultivable hectares (Meadows et al., 1992).

Water represents a second commonsense link between population size and resource use. It is central to the ecological cycles on which we depend and is used by humans for consumption as well as agricultural and energy production. Global water use has tripled since 1950, now standing at between 3,500 and 4,500 cubic kilometers per year, equivalent to eight times the annual flow of the Mississippi River, and representing one-third of the water readily available

---

[8]We say "typically" requires land because of such recent advancements in food production as aquaculture and hydroponics, which are not linked to land resources. It is unlikely, however, that these advancements will soon provide sufficient food production to negate the requirement of land resources.

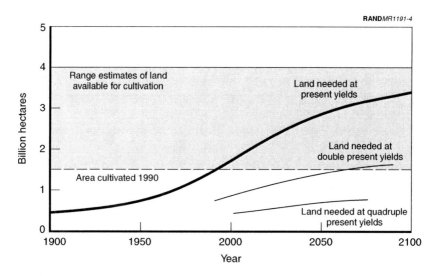

RAND*MR1191-4*

NOTE: The shaded areas show the range of estimates of land that could hypothetically be brought into cultivation, much of which is presently forest land. Population projection based on World Bank forecasts.

SOURCES: Figures adapted from Meadows et al., 1992; land estimates from Higgins et al., 1982, and WRI, 1990; population estimates from Bulatao et al., 1990.

Figure 2.3—Land Required for Food Production

(Goudie and Viles, 1997; Meyer, 1996).[9] Global water consumption rose sixfold between 1900 and 1995, more than double the rate of population growth (WMO, 1997). In the United States, daily per capita water consumption is currently about 185 gallons for domestic tasks (drinking, cooking and washing) (Acreman, 1998).

Human use of environmental resources is nothing new (Wolman, 1993). Even 7,000 years ago, agriculturists and pastoralists altered the lands of the Middle East to increase productivity of the arid environment on which they depended. Beginning in at least 4100 B.C., irrigation water was drawn from the Euphrates, and archaeological data suggest that before 300 A.D. much of the tropical forest of the Maya lowlands in what is now Central America had been cut and species composition consequently affected (Whitmore et al., 1990;

---

[9]4,000 cubic kilometers represents 10^{15} gallons.

Southwick, 1996). What is new, however, is the level of resource use required by a larger-than-ever global population that continues to grow by 80 million annually.

Population size relates not only to the consumption of environmental resources, but also to the environmental pollutants associated with contemporary production and consumption processes. Air, water, and land environments all act as sinks, or repositories, for the pollution generated by production and consumption. The many dimensions of industrial processes make it impossible to generalize about the exact relationship between global population size and pollution. However, researchers have estimated the effect of population size for particular types of pollution in particular locales. Consider air pollution in California. Automobiles, factories, landfills, and airports contribute to local air pollution levels. In a simple sense, population can be related to each of these factors: more people, for instance, means more demand for the consumer goods produced by emission-generating factories. Yet the underlying relationships are not so simple—climate, pollution control legislation, and the technology used to produce goods all combine to determine air quality.

To examine some of these interactions, Cramer (1998) determined county-level associations among emissions, population size, and regulatory efforts in California, and other related factors. Results suggest that a 10 percent increase in population produces an increase in emissions of 7.5 to 8 percent, although population growth has different effects on different types of pollutants (Cramer, 1998).[10] Local population growth is important mainly as a determinant of the volume of consumption. More people, for instance, typically means more cars.[11]

---

[10]In a multivariate statistical analysis, reactive organic gases, oxides of nitrogen (the precursors to ozone), and carbon monoxide were strongly affected by population growth, while oxides of sulfur and small particulate matter were not (Cramer, 1998).

[11]California has a long history with air quality issues, and it provides an excellent example of the important mediating influence of policy on environmental conditions. Since 1960, significant improvements in regional air quality have been made despite a threefold increase in motor vehicle traffic and regional population, combined with substantial industrial development. The improvements are the result of aggressive air quality policies, including an innovative emissions trading program, mediating the impact of population growth on environmental conditions (Switzer and Bryner, 1996). Chapter Four provides further discussion of policies as mediating factors.

Chapter Three

POPULATION DISTRIBUTION: TRENDS AND ENVIRONMENTAL IMPLICATIONS

"Population distribution" refers to the arrangement of population across space, or population's relative geographic location. Population density (population divided by land area) is often used to indicate variation in distribution across regions, and, as such, population distribution is closely related to population size; population density actually represents population size as bounded by a specific locale.

Population distribution has important environmental implications, particularly because many environmental changes are felt locally. As reviewed in this chapter, the environmental implications of population distribution are evidenced by (1) the increased pressure placed on overextended resources in many less-developed nations as a result of relative increases in population densities, (2) the ecological strain put on coastal resources as a result of amenity-driven migration in the United States, and (3) the ecological effects of urbanization, including concentration of pollutants and land-use conversion.

There are many potential policy responses to the environmental implications of local population pressure. Population-oriented policy may aim to reduce local population growth through fertility reduction, thereby lessening pressure on resources. It is important to recognize, however, that political response to the implications of population distribution need not be population-oriented. Rather, policies related to land use, consumption, and production processes have the potential to mitigate localized population-induced environmental change—some through influence on migration patterns, others through influence on production technologies. As examples, localized environmental impact stemming from population distribution and redistribution can be constrained through restrictions on local land use through zoning regulations, designation of conservation areas, or technology mandates in urban industrial regions.

POPULATION DISTRIBUTION TRENDS AND PATTERNS

Changes in population distribution are due to two factors: (1) variations in natural increase that shift the relative proportions of population across locales and (2) migration. The past 40 years have witnessed remarkable changes in the distribution of humans across the globe. In particular, the increasingly urban concentration of population is a prominent contemporary demographic trend.

As for relative distribution of population across the globe, continued high fertility levels in many less-developed regions, coupled with low (or declining) mortality, is resulting in increasingly greater shares of the global population residing in less-developed areas (see Figure 3.1). According to the United Nations, by late 1999, the population of

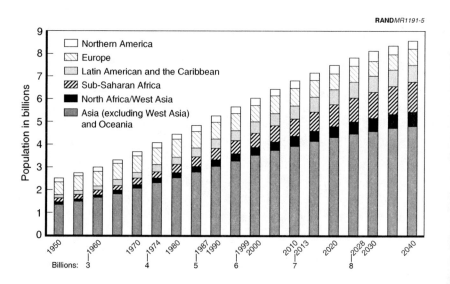

SOURCE: United Nations, 1998c.

Figure 3.1—Regional Distribution of Population

less-developed regions had grown to 4.8 billion, representing 80 percent of the world's population—a 10 percentage point increase since
1960 (UNFPA, 1999b). As a more specific example, Africa's share of
global population is projected to rise to 20 percent in 2050, as compared to only 9 percent in 1960 (UNFPA, 1999b).

Population distribution is also influenced by migration, a complex
process driven by many factors. Individuals can be motivated to
migrate by the "pull" factors of possible destination areas, including
improved employment prospects, the possibility of joining family
members, or other desirable economic or noneconomic amenities.
On the other hand, a lack of employment opportunities at home or
other negative characteristics can act as "push" factors motivating
out-migration (Martin and Widgren, 1996). In recent years, global
transportation and communication have increasingly allowed individuals to respond to the "push" and "pull" forces of migration,
resulting in both migration across national borders (international
migration) and migration within countries (internal migration).

As for international migration, numbers are at an all-time high. The
net flow of international migrants is currently estimated at 2 million
to 4 million annually, and a total of 125 million individuals live outside their country of birth (Martin and Widgren, 1996). The signifi-
cance of internal migration as a force in population redistribution
can be illustrated by U.S. patterns. During 1996–1997, fully 42.1 million Americans moved to a different dwelling unit, representing 16
percent of the population. More than 6 million of these migrants
changed states (Faber, 1998). While migration is often seen as a
response to the changing geography of economic opportunity,
noneconomic motivations are becoming increasingly important. In
fact, within rural areas of the United States, environmental amenities, such as climate, topography, and water-related opportunities,
drive much of rural population change (McGranahan, 1999). Fueled
in large part by this amenity migration, including retirees fleeing cold
winters, 71 percent of the 2,305 rural counties in the United States
gained population between 1990 and 1998 (PRB, 1999a).

Urbanization represents another striking pattern of contemporary
population redistribution. During the past four decades, global
population has experienced a massive urban transition. As recently
as 1960, only one-third of the world's population lived in urban

areas. In 1999, this proportion had increased to nearly half (47 percent, 2.8 billion people) (UNFPA, 1999b).[1] Although some progress has been made in reducing urban fertility levels, natural increase continues to account for about half of urban population growth (UNFPA, 1999b). In addition, the industry and commerce of cities act as magnets to many migrants who are drawn by perceived opportunities and the availability of services. At present, the pace of urbanization is highest in developing regions—the proportion of people in developing countries who live in cities nearly doubled between 1960 and 1990, from less than 22 percent to more than 40 percent (UNFPA, 1999b).

The process of urbanization is projected to continue well into the twenty-first century. Between 1990 and 2025, the number of people who live in urban areas is expected to double, with the vast majority of this urban growth taking place in developing countries (United Nations, 1998c; see Figure 3.2). By 2030, it is expected that nearly 5 billion (61 percent) of the world's people will live in cities (UNFPA, 1999b).

As a result of this high level of urban concentration, several cities have reached unprecedented levels of concentration (see Table 3.1). The number of "megacities" with 10 million or more inhabitants is increasing rapidly, mostly in developing nations. In 1960, only New York and Tokyo had more than 10 million people. By 1999, 17 cities had reached this level of concentration, 13 of them in less-developed regions (UNFPA, 1999b).

THE ENVIRONMENTAL IMPLICATIONS OF POPULATION DISTRIBUTION

Several categories of environmental implications can be related to population distribution and redistribution. First, as less-developed regions cope with an increasing share of global population, pressure will be intensified on already dwindling proximate environmental

---

[1]According to the Population Reference Bureau, "urban" is typically defined as those areas with 2,000 or more inhabitants or national or provincial capitals (1998a).

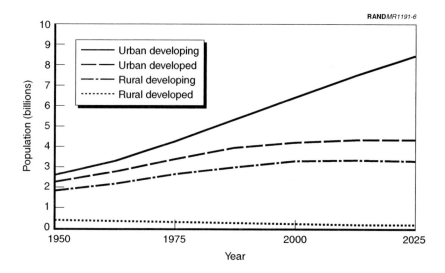

NOTE: Developed regions include North America, Japan, Europe, and Australia and
New Zealand; developing regions include Africa, Asia (excluding Japan), South America, Central America, and Oceania (excluding Australia and New Zealand). The European successor states of the former Soviet Union are classified as developed regions,
while the Asian successor states are classified as developing regions.
SOURCE: United Nations, 1998c.

Figure 3.2—Urban Population Growth, 1950–2025

resources. Second, the redistribution of population through migration also brings shifts in relative human-induced environmental
pressures—perhaps easing localized environmental change in some
areas while increasing such change in others. Finally, rapid urbanization particularly in less-developed regions simply outpaces the
development of infrastructure and environmental regulations, often
resulting in high levels of air and water pollution. Each of these relationships is discussed below.

As for resource pressure, many areas in less-developed regions are
already facing shortages of arable land, clean water, and sufficient
fuelwood; increases in local population densities will likely exacerbate these scarcities. In the early 1980s, wood supplied the vast
majority of household energy for domestic cooking and heating in
many less-developed regions—82 percent in Nigeria, 92 percent in

Table 3.1

The World's Twenty-Five Largest Cities, 1995

	Population (millions)	Average Annual Growth Rate 1990–1995 (percent)
Tokyo, Japan	26.8	1.4
Sao Paulo, Brazil	16.4	2.0
New York, U.S.	16.3	0.3
Mexico City, Mexico	15.6	0.7
Bombay, India	15.1	4.2
Shanghai, China	15.1	2.3
Los Angeles, U.S.	12.4	1.6
Beijing, China	12.4	2.6
Calcutta, India	11.7	1.7
Seoul, Republic of Korea	11.6	2.0
Jakarta, Indonesia	11.5	4.4
Buenos Aires, Argentina	11.0	0.7
Tianjin, China	10.7	2.9
Osaka, Japan	10.6	0.2
Lagos, Nigeria	10.3	5.7
Rio de Janeiro, Brazil	9.9	0.8
Delhi, India	9.9	3.8
Karachi, Pakistan	9.9	4.3
Cairo, Egypt	9.7	2.2
Paris, France	9.5	0.3
Metro Manila, the Philippines	9.3	3.1
Moscow, Russian Federation	9.2	0.4
Dhaka, Bangladesh	7.8	5.7
Istanbul, Turkey	7.8	3.7
Lima, Peru	7.5	2.8

SOURCE: UN Population Division, 1995.

Tanzania, and 94 percent in Nepal. Even with current population pressure, the fuelwood demand in many countries simply outpaces sustainable supplies—wood is being cut faster than it can be replenished through natural growth. Consumption exceeds sustainable supply by 70 percent in Sudan, by 150 percent in Ethiopia, and by 200 percent in Niger. Increasing population densities result in even less supply per capita, and although other factors play a role in wood shortages (e.g., failure to encourage afforestation and/or the use of alternative energy sources), nearly 75 percent of the increase in wood demand from 1980–2000 was estimated to be caused by local population growth (see Figure 3.3; UNFPA, 1991, p. 49).

Arable land resources also feel the squeeze of population pressure. In rural Guatemala, when high fertility levels and falling mortality rates increased local population densities, the need for agricultural land intensified. Two primary responses emerged: fragmentation of land resources as small family holdings were divided among heirs and out-migration, often resulting in deforestation of other rural areas to expand agricultural production. The results: increasingly smaller agricultural holdings, some too small to provide sufficient subsistence production, in conjunction with high levels of deforestation—between 1950 and the mid-1980s, roughly half of the forested area was cleared. In this case, heightened population pressures, in conjunction with land tenure policies, brought about land-use changes in response to scarcity, changes that further endangered dwindling land and forest resources (Bilsborrow and Stupp, 1997).

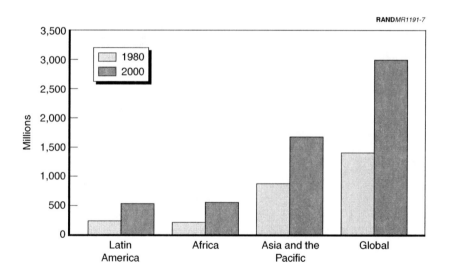

SOURCE: Adapted from UNFPA, 1991, p. 48.

Figure 3.3—Population Experiencing Fuelwood Deficit, 1980 and 2000

As reviewed in Chapter Six, related scenarios play out in other regions of the world.[2]

Population redistribution through migration can affect environmental conditions, particularly when high levels of demographic pressure are exerted on fragile ecosystems. The coastal zones of the United States are an example. Although coastal counties (excluding Alaska) constitute only 11 percent of U.S. land area, they are home to 53 percent of the population (Culliton et al., 1990; Culliton, 1998). Coastal population is expected to reach 165 million by 2015, an average daily increase of 3,600 people (Culliton, 1998). California, Florida, Texas, and New York consistently account for a significant portion of coastal population growth. Yet, recent increases in migration fueled by recreation and retirement are also bringing rapid growth to the barrier islands of the Mid-Atlantic states and parts of the Gulf Coast (Frey, 1995). From 1980 to 1990, Florida counties experienced net growth of up to 781 percent (to 294 persons per square kilometer), counties of the central Atlantic coastal barriers had growth rates of up to 300 percent (to 294 persons per square kilometer), and Gulf Coast counties associated with the barriers of Louisiana and Texas had growth rates up to 155 percent (to 47 persons per square kilometer) (Bartlett, Mageean, and O'Connor, 1999). In North Carolina, the narrow coastal islands of Bodie and Hatteras absorbed most of Dare County's 280 percent growth, adding an additional 16 persons per square kilometer (Bartlett, Mageean, and O'Connor, 1999).

These increases in coastal population densities bring reduced vegetation cover, habitat loss, and resulting declines in species diversity (McAtee and Drawe, 1981). Greater levels of human activity in coastal areas also result in other significant ecological changes, such as declining beach elevation and changes in soil pH and average soil temperature—all having ultimate impacts on ecosystem sustainability (McAtee and Drawe, 1981). Figure 3.4 illustrates where high

---

[2]For additional case studies demonstrating land-use changes resulting from land shortages see DeWalt, Stonich, and Hamilton, 1993, May 1995, and Bilsborrow and Hogan, 1999. For discussions of the role of water shortages in population-landscape interactions, see various publications by Falkenmark and colleagues (e.g., Falkenmark, 1991, 1994, Falkenmark and Suprapto, 1992; Falkenmark and Widstrand, 1992). The importance of accessible fuelwood is discussed in DasGupta, 1995.

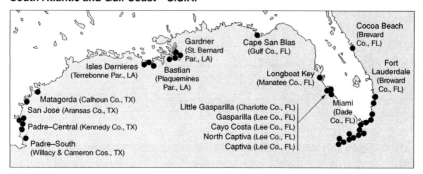

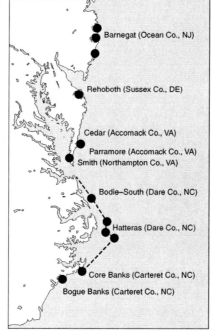

SOURCE: Bartlett, Mageean, and O'Connor, 2000.

Figure 3.4—Coastal Barrier Dune Ecosystems Experiencing High Levels of Population Change, Central and South Atlantic and Gulf Coasts, United States

levels of population pressure overlap with coastal barrier dune ecosystems along the South Atlantic, Central Atlantic, and Gulf Coasts of the United States, therefore putting these areas at risk of human-induced ecological decline.

The final aspect of population distribution to be related to the environment is urbanization, the environmental implications of which can be considered either positive or negative depending on which particular impact is being examined (Pebley, 1998). On the positive side, cities promote efficiencies in transportation, housing, utilities, distribution of goods, and provision of services (Southwick, 1996). In addition, the recycling of inorganic materials can be easier in cities because the population is concentrated (Qutub, 1992). Finally, assuming urban sprawl is controlled, high-density settlements can help preserve natural habitat outside of urban areas. Imagine, for instance, if all the urban dwellers in the world were scattered over the landscape at very low densities (Southwick, 1996).

On the negative side, however, at least four general areas of environmental consequences result from the high population densities accompanying urban development. First, the waste produced by such densities is beyond that readily absorbed by the surrounding environment, resulting in high concentrations of pollutants (Benneh, 1994). The high levels of air pollution characterizing many megacities testifies to the inability of the environment to absorb the wastes produced by high densities of consumers and production processes (see Table 3.2; also see Brennan, 1996).

Second, the rapid pace of urban growth occurring in developing regions, in addition to the sheer size of megacities, greatly hinders the development of adequate infrastructure or regulatory mechanisms to handle the environmental impacts of human concentration. As an example, rapidly increasing population densities completely overwhelmed the sewage system in Karachi, Pakistan (population 10 million), often operating at only 15 percent capacity as a result of breakdowns and clogged pipes. Much of the sewage eventually contaminated drinking-water wells because it had leaked into the surrounding soil (Rahman, 1995). Such contamination is responsible for many waterborne diseases, including diarrheal disease, cholera, typhoid, and hepatitis A and E. Especially in developing regions,

Table 3.2

Status of Pollutants in the Megacities, 1992

City	SO_2	SPM	Pb	CO	NO_2	O_3
Bangkok	○	★	●	○	○	○
Beijing	★	★	○	—	○	●
Bombay	○	★	○	○	○	—
Buenos Aires	—	●	○	★	—	—
Cairo	—	★	★	●	—	—
Calcutta	○	★	○	—	○	—
Delhi	○	★	○	○	○	—
Jakarta	○	★	●	●	○	●
Karachi	○	★	★	—	—	—
London	○	○	○	●	○	○
Los Angeles	○	●	○	●	●	★
Manila	○	★	●	—	—	—
Mexico City	★	★	●	★	●	★
Moscow	—	●	○	●	●	—
New York	○	○	○	●	○	●
Rio de Janeiro	●	●	○	○	—	—
São Paulo	○	●	○	●	●	★
Seoul	★	★	○	○	○	○
Shanghai	●	★	—	—	—	—
Tokyo	○	○	—	○	○	★

RAND*MR1191-T2*

★ Serious problem, WHO guidelines exceeded by more than a factor of two

● Moderate to heavy pollution, WHO guidelines exceeded by up to a factor of two (short-term guidelines exceeded on a regular basis at certain times)

○ Low pollution, WHO guidelines normally met (short-term guidelines are exceeded occasionally)

— No data available or insufficient data for assessment

NOTE: SO_2 is sulfur dioxide; SPM is suspended particulate matter; Pb is lead, CO is carbon monoxide, NO_2 is nitrous oxide, and O_3 is ozone.

SOURCES: WRI, 1994, p. 198.

many waterborne diseases are the principal causes of infant and child mortality. For example, diarrheal disease, the major waterborne disease, ranks as the leading cause of morbidity in the world and is estimated to be responsible for over 3 million child deaths in 1990 (WHO, 1998c).

Third, urbanization often results in alteration of local climate patterns. Concentrations of artificial surfaces, such as brick and concrete, replace natural ground and alter heat exchange patterns, thereby creating "heat islands." In cities with more than 10 million people, the mean annual minimum temperature can be as much as 4 degrees Fahrenheit higher than in nearby rural areas, and these changes can affect climate, water flows, and plant and animal diversity, as well as human health (Berry, 1990).

Finally, poorly planned urban development can result in significant conversion of land from habitat or other purposes, such as agriculture. The consequences of such development patterns are especially apparent in the extended metropolitan regions resulting from urban sprawl in the United States. During 1992–1997, 16 million acres of forest, cropland, and open space were converted to urban and other uses, representing an 18 percent increase in the nation's developed land area (USDA, 1999a).[3] The land developed during this five-year period was greater than the total land developed during the 10-year period 1982–1992 (13 million acres), suggesting an increase in the pace of sprawling development (USDA, 1999a). In the Chesapeake Bay area, development pressures have reduced tree canopy from 51 percent to 37 percent in just the past 25 years (USDA, 1999b). In California, much prime agricultural land is adjacent to rapidly expanding urban areas. About 250,000 acres (4.5 percent of California cropland) were lost to development during the period 1982–1992 (USDA, 1999b).[4] Watersheds in the San Joaquin Valley have been especially affected, with the level of land conversion ranking among the top 2 percent of the nation's more than 2,100 watersheds (USDA, 1999b).[5]

---

[3]The developed land category includes large tracts of urban and built-up land; small tracts of built-up land, less than 10 acres in size; and land outside of these built-up areas that is in roads, railroads, and associated rights-of-way (USDA, 1999).

[4]As stated previously, the density of urban areas can also be considered positive in the sense of preserving habitat. The difference here is one of planning and appropriate infrastructure development.

[5]Watersheds are defined as U.S. Geological Survey Hydrologic Cataloging Units (8-digit) (USDA, 1999b).

Chapter Four

POPULATION COMPOSITION: TRENDS AND ENVIRONMENTAL IMPLICATIONS

"Population composition" refers to the characteristics of a particular group of people. These characteristics include the distribution of a population across age categories and the number of men relative to the number of women. Age and sex composition are the most often considered aspects of population composition, although socio-economic characteristics, such as household income levels, can also be considered compositional. Income is included here because of its considerable relationship with environmental factors.

While the suggested relationships discussed below between age composition and environmental factors are largely speculative, they are based on existing research linking this composition characteristic to such processes as migration. As such, the potential environmental effects of changes in composition represent areas relevant for further investigation with regard to the relationship between population and the environment. The relationship between income and environmental factors is better understood, with evidence suggesting that consumers' aspirations rise with income.

As for policy response with regard to these environmental implications, relevant measures tend to fall outside the realm of population policy per se primarily because of our lack of adequate understanding of the environmental implications of population composition. As examples of policy responses related to population composition, rural economic development policies could stem the tide of young rural-to-urban migrants in search of employment opportunities, while consumer education might heighten awareness of the environmental costs of energy-intensive, income-driven consumption patterns.

POPULATION COMPOSITION TRENDS AND PATTERNS

As for age composition, the legacy of changing fertility and mortality levels can be seen in the distribution across age groups within the remaining population. Today's global population is characterized by the largest-ever generation of young people on every continent except Europe, the product of recent high fertility levels. Although the relative proportion of individuals in this age group is on the decline (because of aging populations as discussed below), the absolute numbers of individuals between the ages of 15 and 24 continues to rise. Currently, 1.05 billion individuals are between the ages of 15 and 24, and this age group is rapidly expanding in many countries (UNFPA, 1999a). In contrast, in 1985, 769 million individuals were aged 15–24 years.

At the same time, declining mortality and increased longevity have resulted in the expansion of older populations. Worldwide, the average life expectancy in 1950 was 46 years; in 2050, the average life expectancy is projected to be 76 years (UNFPA, 1999b). These changes in life expectancies are reflected in the increasing percentage of population in older age categories. In the developed regions, 7.9 percent of the population was 65 years and older in 1950; by 2050, 24.7 percent of the population is expected to be within this age group. Worldwide, the number of older persons being added to the world's population is approaching 9 million per year and is expected to grow to about 14.5 million per year in 2010–2015 (UNFPA, 1999a). The relationships between changing fertility and mortality patterns and age composition are demonstrated by Figure 4.1.

Age composition has important implications for future population growth because younger populations possess greater growth momentum. In other words, even if fertility were to immediately fall to replacement levels of approximately two children per woman in all nations, population size would continue to increase because of the large number of women in reproductive years. On the other hand, lower levels of population growth would be expected for popu-

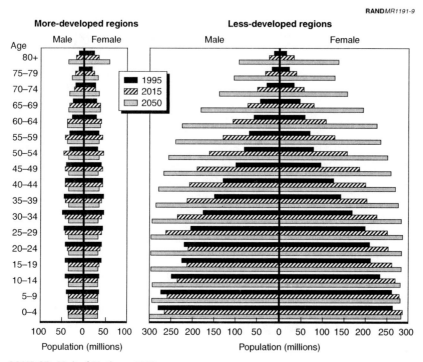

SOURCE: United Nations, 1998.

Figure 4.1—Age and Sex Structure of the Population, 1995, 2015, and 2050

lations in which older women constituted a relatively larger share.[1] Because of the young age composition of today's population, over the next 25 years, population momentum is expected to account for almost three-fourths of population growth in developing countries and nearly all population growth in East Asia (Bulatao, 1998). In addition, although the percentage of individuals aged 15–24 will decline over the next several decades, the actual number in this age group will continue to rise as a result of population momentum (769 million in 1985, 863 million in 1995, and 1.16 billion in 2050) (United Nations, 1999).

---

[1] See Bulatao (1998, pp. 4–6) for a useful description of population momentum.

A population's income profile can be considered another compositional characteristic—one with many different implications, and one which reflects many different social and economic processes. Poverty reduction has occurred in some regions of the developing world; the percentage of individuals living on less than one U.S. dollar per day (a frequently used measure of poverty) is declining (World Bank, 1999). Also suggesting improvements in income levels, between 1980–1997, per capita private consumption grew annually by an average of 2.7 percent in low-income nations, compared to 1.2 percent in high-income countries (World Bank, 1999).[2] Improvements in income are less positive, however, when examined from the perspective of absolute numbers of individuals. Although poverty percentages are on the decline, the *number* of individuals living on less than $1 per day continues to rise as a result of population growth. In 1987, 1.2 billion individuals were living below this level, 1.5 billion in 1999, and if present trends continue, 1.9 billion are projected to live on less than $1 per day by 2015 (World Bank, 1999).

POPULATION COMPOSITION AND THE ENVIRONMENT

Consideration of demographic factors, such as age composition and income profiles, can refine our understanding of the *mechanisms* through which population relates to environmental conditions.

For example, migration propensities vary by age, with young adults exhibiting the highest likelihood of moving (Long, 1988). Much of this movement is fueled by education and employment prospects, as individuals leave the parental home in search of new opportunities. Given the relatively large younger generation characterizing many developing nations, we might anticipate increasing levels of urban-

---

[2]These figures represent measurements of private consumption corrected for the distribution of income within a nation (see Technical Notes, World Bank, 1999). The rate of poverty reduction is, on average, proportional to the distribution-corrected rate of growth in private consumption. Private consumption is the market value of all goods and services, including durable products (such as cars, washing machines, and home computers), purchased or received as income in kind by households and nonprofit institutions.

Income classifications are as defined by the World Bank (1999). Low-income countries are those with 1998 per capita GNP of $760 or less. High-income countries are those with 1998 per capita GNP of $9,361 or more.

ization as these individuals experience the "pull" of urban opportunity. As such, consideration of what age composition might portend for the demographic future can provide insight into prospective urbanization patterns and related environmental concerns.

Varying migration propensities relate to another potential environmental implication of age composition: levels of retirement migration can be expected to increase as populations age, particularly in cultures that exhibit these life cycle–related patterns of movement. In the United States, for instance, permanent or seasonal migration to areas with environmental amenities may increase, as may use of park and recreation areas in these and other areas (Orians and Skumanich, 1995).

Income is especially important when it comes to considering the environmental implications of population composition. The level of development and per capita income exhibit an inverted U-shaped relationship (also known as a Kuznets curve) with certain types of environmental pressures, particularly industrial emissions (UNEP, 1997; see Figure 5.4, p. 43). According to this general framework, emissions are minimal in regions characterized by low levels of development and income, yet high in the middle-development range as economies move through the early stages of industrialization. At more advanced industrial stages, however, some environmental pressures ease because (1) "dirty" economic sectors, such as heavy manufacturing, become less important than "clean" economic sectors, such as many service industries; (2) very high income consumers use fewer environmentally degrading technologies; (3) rising incomes are associated with a greater demand for environmental quality; and (4) resources are limited (Perrings, 1998). Also related to development stage and environment, an interesting distinction between the types of environmental pressures typifying different development levels has been outlined by the World Health Organization (WHO, 1998a). Traditional environmental hazards include lack of access to safe drinking water, inadequate basic sanitation, and indoor air pollution. As economies develop, these traditional hazards decline, but more technologically driven forms of water and air pollution take their place, ultimately threatening to cause climate change and ozone depletion.

The Kuznets curve does not, however, entirely capture the complexity of the relationships among population, development, income, and environmental conditions. At the local level, population pressures and poverty can interact to result in unsustainable use of proximate resources to meet short-term livelihood needs. Here then, lower levels of development and income are associated with greater environmental pressure at the local scale. It is argued that poverty can induce people to behave as if they are myopic, making use of local resources without consideration of global effects or the implications for future generations (Perrings, 1998). As described next, research in the Philippines illustrates some of these important interactions.

The Philippines' population growth of 2 percent per year resulted in increases in population density from about 190 persons per square kilometer during the late 1980s to more than 210 per square kilometer in 1990, a 10 percent density increase in less than five years. It is now one of the most densely populated countries in the world. In addition, more than half of the population lives below the poverty line. These high levels of poverty and population pressures interact to increase the demand for agricultural and fuelwood resources. In turn, these resource shortages act as "push" factors as poverty-stricken migrants relocate into frontier upland forests. As a result, more than 30 percent of total cultivated area was upland forest in 1987, compared to only 10 percent in 1957 (Cruz et al., 1992; Cruz, 1997). The environmental implications of population pressures and poverty have also been studied in Mexico, with results suggesting, again, that poverty is an important factor in deforestation (Deininger and Minten, 1999).

On the other end of the development spectrum, today's very rich are criticized for consuming disproportionate amounts of energy and producing disproportionate amounts of waste. As related to the Kuznets curve, the negative correlation between income and environmental pressure suggests that contemporary development stages have yet to reach the point of being environmentally benign. Rising incomes have environmental implications stemming from their association with changing production and consumption patterns. As incomes rise, consumer aspirations change and meeting heightened consumer demand typically means rising pollution levels, at least within the confines of present production processes. In other words, higher socioeconomic status also carries detrimental environmental effects. A positive linear correlation has been demonstrated between per capita income and carbon dioxide (CO_2) emissions at the national level, where each $1 increase in per capita GDP is associated with a 1.4 kilogram increase in per capita CO_2 emissions (Perrings, 1998). As a city-specific example, in Bangkok, an increase of 1,000 baht ($26) in monthly income equates with a 0.1 kilogram increase in vehicle-related air emissions per year (de Souza, 1999). In general, income exhibits an important correlation with environmental change, although the direction varies according to the environmental aspect under consideration.

Chapter Five

MEDIATING FACTORS: SCIENCE AND TECHNOLOGY, INSTITUTIONS AND POLICY, AND CULTURE

The previous chapters have reviewed several dimensions of the relationship between population and the environment, focusing on various demographic factors and offering examples of their relation to select aspects of the environment. The relationship between population and the environment is, however, even more complex than this focus would suggest. Aspects of society relating to technologies, institutions, and culture, alter the ways in which demographic and environmental factors interact. These "mediating factors" are reviewed here, with brief examples provided where useful. Again, the complexity of these relationships prohibits a comprehensive review in such limited space, so the following discussion is designed to illustrate the factors that shape the environmental implications of demographics.

SCIENTIFIC AND TECHNOLOGICAL FACTORS

Scientific and technological advancements have mediated the relationship between population and the environment since prehistoric time. The scope of their impacts can be demonstrated by four simple examples: the discovery of the usefulness of fire brought about more stable prehistoric societies; agricultural processes were greatly enhanced by the development of technology allowing the harnessing of draft animal energy; advancements allowing the use of wind for sea transport enabled new migration opportunities, and; technology harnessing the energy offered by coal and electricity allowed for the expansion of urban centers (Boserup, 1981; Colombo, 1996). In

35

some cases, these scientific advancements and technological changes resulted in environmental modification beyond what would be anticipated stemming exclusively from demographic factors (e.g., the pace of land cover change was determined not only by population size, but also by agricultural technology). In other cases, these advancements allowed for shifts in the demographic factors that modified the environment (e.g., the scale of population redistribution afforded by advancements in travel).

During the past 100 years, improvements in agricultural technology, such as mechanized cultivation, high-yield seed varieties, fertilizers, and pesticides, have increased the yield per acre by more than 100 percent (Rasmuson and Zetterstrom, 1992). These advancements enabled the sustenance of the world's growing population—a scenario not considered by Malthus (1798) in his predictions of widespread starvation. In fact, some argue that since humans bring about such innovation, population itself is not necessarily detrimental to the environment. Rather, humans may be the ultimate resource in coping with environmental change because greater numbers of people mean more creativity and innovation (Simon 1981, 1996).

Several studies support the perspective that high levels of demographic pressure may induce innovation or at least intensify the application of existing technologies. In northern Nigeria, population densities are high and the majority of land is already under cultivation. Those with small landholdings have made significant investments in land improvement through sustainable management of soil fertility and increased use of farm trees. In areas lacking intense population pressure, such innovations had not yet been adopted, suggesting that the resource pressure spurred changes in land management (Mortimore, 1993).[1] In another study conducted at the national level using data reflecting population and environmental conditions in 85 developing nations, a positive relationship was found between rural population growth and intensification of irrigation and fertilizer use (Bilsborrow, 1992).

---

[1] In this study of the association between demographic pressure and innovation (Mortimore 1993), causality is *inferred* as opposed to specifically tested.

The technological changes that have had the most effect on environmental conditions relate to the ways in which humans use energy. In fact, energy use is one of the primary links between population and the environment. Transitions in technologies related to energy represent an important aspect of the modernization process, as reflected in the change since 1850 in energy consumption (see Figure 5.1). In particular, the consumption of oil, coal, and natural gas has significantly increased during the past century.

Until about 1960, the developed nations were responsible for most of the increases in energy consumption. Prior to that time, consumption in the less-developed regions consisted of relatively small amounts of wood and traditional biomass materials such as cow dung (Weyant and Yanigisawa, 1998). Since the 1960s, however, many less-developed nations have experienced some level of industrial development, resulting in increased reliance on resource-intensive and highly polluting energy production processes. The People's Republic of China (PRC) provides an example of the links between population size, composition, and energy use. The PRC's population

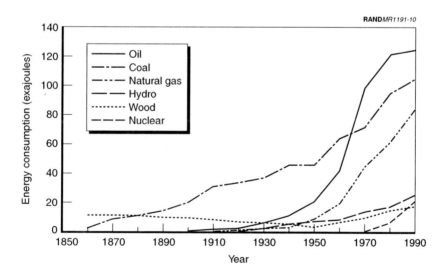

SOURCE: Weyant and Yanigisawa, 1998, p. 208.

Figure 5.1—World Energy Consumption, 1850–1990

currently exceeds 1 billion, and as development proceeds, per capita income is expected to rise dramatically. Given the nation's large coal reserves and changing consumer aspirations arising from income increases, carbon emissions are expected to rise to the point that the PRC becomes the world's largest producer of emissions within the next 20–30 years (Weyant and Yanigisawa, 1998). Obviously, improved efficiencies in energy production could greatly diminish the environmental impact expected from the interaction between China's population size and changing consumption patterns.

INSTITUTIONAL FACTORS, THE POLICY CONTEXT

Institutional response represents a significant mechanism through which humans react to environmental change. In particular, policy plays a key role in determining the ultimate effect of humans on the environment. Following are four examples. The first represents a positive policy outcome operating at the global scale, and the other three demonstrate the potential negative effects of misguided policy.

Example 1, the Montreal Protocol of 1987

During the 1980s, evidence began to suggest that the Earth's protective atmospheric ozone layer was being significantly eroded. The ozone layer shields humans from potential eye damage and skin cancers caused by the sun's high-energy ultraviolet radiation. It also protects crops and livestock (Meyer, 1996). The primary culprit in ozone depletion appeared to be human-induced—the production of synthetic organic compounds, chlorofluorocarbons (CFCs), used in refrigeration, solvents, and propellants. Although ozone decline had human origins, the rate of ozone depletion was lessened during the 1990s, even in the face of continued world population increases. Here, the relationship between population and the environment was tempered by the ratification of the Montreal Protocol in 1987 (see Figure 5.2). As a result of aggressive campaigning by international nongovernmental organizations (NGOs), and political leadership from the United States, several European countries, and the United Nations Environment Program (UNEP), the unprecedented international agreement aimed to reduce and eventually eliminate the emissions of manufactured ozone-depleting substances (Jasanoff

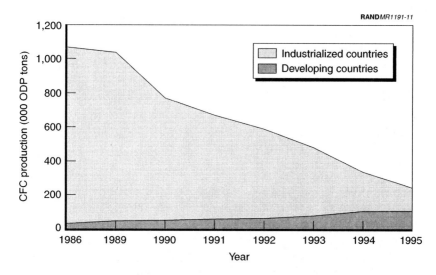

NOTE: Ozone-depleting potential (ODP) tons is a measure by which ozone-depleting substances are weighted according to their ability to destroy ozone.

SOURCES: WRI, 1998, p. 1777; Oberthür, 1997, p. 30.

Figure 5.2—Annual Production of CFCs, 1986–1995

and Wynne, 1998). Under current agreements, CFC consumption has dropped more than 70 percent (WRI, 1998), with the ozone layer expected to return to normal by the middle of the next century (UNEP, 1998).

As demonstrated by this example, many factors have combined to determine the scale, scope, and intensity of ozone depletion. Yet demographic factors remain important. In particular, population influenced environmental conditions by providing a market for destructive technologies. To look at this relationship another way, the technology allowing CFC production operated as a mediating influence between population size and the environment. In this instance, however, it was the policy response that ultimately defined the relationship between technology, consumption, population, and environmental change.

Example 2, the Aral Sea Basin

Policy actions not only ameliorate environmental decline, but on occasion misguided policy may also become a powerful force behind degradation. The ecological and social dilemmas facing the Aral Sea basin offer an extreme example of the effects of policies regarding resource use. This basin in central Asia is shared by several nations, mainly Uzbekistan and Kazakhstan (Glazovsky, 1995; Micklin, 1997). Since 1960, the Aral Sea has shrunk 40 percent (see Figure 5.3) and, while a component of the decline stems from natural variations, research has demonstrated that human forces are the primary cause behind the ecological destruction. In particular, irrigation policies of the former Soviet Union appear to account for this trend. Canals were built to draw water from the Amu Darya and Syr Darya river basins to support agricultural development in the region. Today, the dried sea floor has changed the original coastline and altered the local precipitation cycle, 20 of 24 native fish species have disappeared, and the number of bird species has decreased from 319 to 168 (Goudie and Viles, 1997; Micklin, 1997). The expansion of agriculture has also heightened chemical pollution in the area. All of these factors combine to feed back into human processes—the vast majority of fisheries have closed resulting in a lack of local economic opportunity and infant mortality runs extremely high (110 infant deaths per 1,000 live births in some areas) (Goudie and Viles, 1997; Micklin, 1997).

Example 3, Mono Lake, California

The complex interactions between demographic pressures, environmental context, and institutional response have also played out in California. In 1941, the Los Angeles Department of Water and Power began diverting water from the Sierra Nevada's Mono Lake basin to meet the needs of southern California's burgeoning population. As a result of the diversion, the lake's surface began dropping by approximately one foot per year. By 1955, the lake's surface had dropped 12 feet; by 1995, more than 40 vertical feet had been lost. With the decline in water volume, sandy beaches became sticky mud, losing much of their recreational appeal. In addition, sediment flows were changed, lagoons and wetlands altered, and strange natural

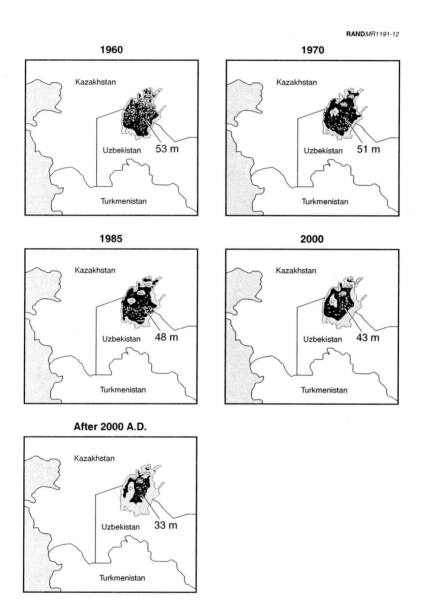

SOURCE: Modified after Hollis (1978) as presented in Goudie and Viles, 1997.

Figure 5.3—Aral Sea, 1960–2000

sculptures known as tufas were exposed. Water plants vanished, affecting naturally occurring brine shrimp and alkali fly populations. This, in turn, resulted in a decline of migrating bird populations dependent on the lake's aquatic life. In sum, the lake's ecosystem was severely disrupted (Hart, 1996).

Litigation on behalf of Mono Lake began in 1979, initially promoted by the National Audubon Society and the grassroots Mono Lake Committee. Years of controversy preceded the landmark decision in 1994 by the State Water Resources Control Board, which amended Los Angeles Department of Water and Power's diversion rights in the Mono basin. The legislation calls for a lake level of 6,392 feet above sea level, which will take about 20 years to achieve but allowing limited diversions for municipal water supplies. Since 1994, additional agreements have been developed requiring stream and waterfowl habitat restoration (MLC, 1999).

Example 4, Energy Policy in India

The critical importance of timely organizational response in mediating the relationship between population and the environment can also be seen in India, a nation whose population grew from 548 million to 846 million between 1971 and 1991, an increase of more than 50 percent in 20 years. A study examining population growth, poverty, and environment during this period concludes that some of the dilemmas facing India today are not specifically the result of population growth or resource shortages per se, but rather of the insufficient institutional and policy response necessary to mediate these relationships. For instance, persistent shortages of coal and coal-based electricity did not result from resource shortages or inadequate technologies but from a failed energy policy. Among other factors, inefficient public sector monopolies continued to operate, and prices were set too low to rein in demand (Repetto, 1997).

Many other examples of the important mediating influence of institutions and policy could be offered. Consider, for instance, the creation of the U.S. Environmental Protection Agency in 1970 and the subsequent implementation of the Clean Air Act (1970), the Safe Drinking Water Act (1984), and regulations regarding hazardous waste disposal (e.g., Resource Conservation and Recovery Act of 1976).

The mediating influences of both technology and policy vary in important ways across the development continuum (World Bank, 1992). The processes of industrialization and modernization that characterize the development of today's developed nations, entail environmentally unsound production processes. The development also often occurred in contexts lacking sound environmental regulation. As reflected in Figure 5.4, these early development stages brought increasing environmental pressures as a result of the adoption of energy-intense production and consumption patterns and the lack of pollution-control regulatory mechanisms. As economies move along the development continuum, however, pollution controls tend to be implemented, along with advances in production, and eventually the adoption of consumption patterns that yield less environmental pressure.

Concerns have arisen about the application of similar development processes to the world's expanding economies. Given that more than 90 percent of future population growth is expected to occur in

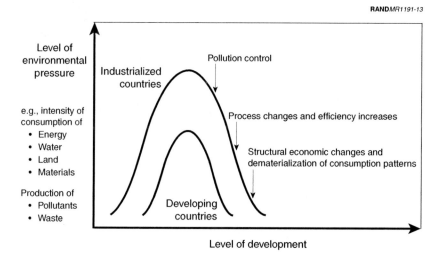

SOURCE: UNEP, 1997, p. 3.

Figure 5.4—Conceptual Representation of the Relationship Between Economic Development and Environmental Pressure

developing nations, consumption-driven environmental pressure in these regions could portend environmental disaster. However, some evidence suggests it is possible to reduce the environmental implications of economic development: the rate of degradation in some of today's developing nations has been slower than that experienced by industrial countries when they were at similar stages of economic development (UNEP, 1997).

CULTURAL FACTORS

Cultural factors encompass the meanings and ways of life that define a society, including beliefs, values, norms, traditions, and symbols (Mooney, Knox, and Schacht, 1997). Here we provide examples of how such cultural factors as gender roles and societal perceptions of natural environments influence the ways in which demographic factors are brought to bear on environmental context.

With regard to gender roles, in the social context of many developing nations, women are the managers of the daily living environment, responsible for the collection of resources necessary for household maintenance. And yet, while primarily responsible for resource collection, women often have less security in access to these resources compared to men. In Zimbabwe, for instance, girls have no rights of inheritance from their parents, and women acquire access to land only through their fathers, husbands, and brothers. This insecurity of tenure influences women's relationship with the environment. As a specific manifestation, women are significantly less likely to plant trees for food, medicine, and fuelwood in areas where future access is uncertain (Fortmann, Antinori, and Nabane, 1997). In societies where women are restricted access to land ownership, they tend to also be ineligible for credit, cooperative membership, and programs designed for innovative land management approaches. In both of these cases, gender roles mediate the relationship between population and the environment by influencing resource management strategies.

Nature-society relations offer another example of cultural factors' mediating influence. For example, distinctive patterns have been shown to exist with regard to attitudes, knowledge, and behavior toward wildlife and conservation across three industrial democracies: the United States, Japan, and Germany. While Americans and

Germans express a broad appreciation for a variety of animals, Japanese culture places greater emphasis on the experience of nature in controlled, confined, and highly idealized circumstances (e.g., bonsai, rock gardening, flower arranging) (Kellert, 1991, 1993). These cultural variations in perceptions of wildlife influence conservation strategies because public support for various policy interventions will reflect societal values (Kellert, 1985).

Chapter Six

SPECIFIC ARENAS OF INTERACTION: CLIMATE CHANGE AND LAND-USE CHANGE

As suggested in the previous chapters, generalizations about the relationship between population and the environment are difficult because of the many types of demographic factors, multiple facets of the environment, and various mediating influences acting on the relationship. To provide more specific details with regard to certain relationships, this chapter brings together the information provided thus far to tell the story of the relationship between population and the environment with regard to two specific issues: climate change and land-use change.

Many other aspects of the environment are influenced by demographic factors but are not discussed here (e.g., air quality, water quality and quantity). Four reasons underlie the choice of climate change and land-use change as specific examples: (1) relatively clear scientific evidence exists with regard to the role of demographic factors for these two topics; (2) each demonstrates how the interaction of population and the environment operates at a different spatial scale—while climate change is a global issue, land-use change operates at the local and national scales; (3) different demographic perspectives are represented, and while climate change is primarily shaped by population size and growth, changes in land-use often arise from shifts in population distribution; and (4) several types of mediating influences are also represented—population is not the only force brought to bear on climate change and land-use change because technology, political issues, and cultural factors all play critical roles in determining the ultimate nature of the relationships.

The examples here also demonstrate the difficulties inherent in esti-mating the influence of demographic factors on the environment. Several methodological issues arise. For example, data on ecological systems are not often easily merged with data on social systems. Here, again, climate change can provide an example. Climate mod-els operate on a macro scale, examining patterns of atmospheric and ocean circulation at the global level. The linking of detailed demo-graphic data to such large-scale models is problematic because of both data comparability and the scale of analysis. There is also the issue of inferring causality: While research may uncover a positive *relationship* between demographic factors and a particular form of environmental change, association does not equate with causation.

Other factors, perhaps reflected by population processes, may actu-ally be the primary driving forces. For example, consider the "push" factors acting on rural-to-urban migrants in many developing regions. In these contexts, migration is a population process that can exacerbate urban environmental ills; yet the process itself may be indicative of rural land shortages or inequitable land tenure systems. Therefore, while a relationship may be apparent between migration and urban environmental pressures, more intricate interrelations among demographic, environmental, and organizational processes are at work.

Including adequate representation and measurement of relevant mediating factors is a related methodological difficulty. No agree-ment exists, for instance, on how best to empirically capture the impact of international markets on land-use decisions made by Amazonian farmers.

Finally, all of these concerns relate to the question of appropriate analytical units—the influence of national policies may best be rep-resented by examination of environmental outcomes across coun-tries. Land management decisions, however, take place at the household level.

The following topical sections summarize research findings regard-ing the relationship between demographic factors, climate change, and land-use change. In the process, we provide examples of the ways in which researchers have attempted to cope with these methodological difficulties.

POPULATION AND CLIMATE CHANGE

In a general sense, climate change refers to alterations in the nature of the Earth's typical weather patterns. Such changes can be found throughout history. For example, a relatively warm period occurred in Europe during the Middle Ages between 1000 and 1400 A.D., succeeded by a Little Ice Age from 1400 to 1800 A.D. (Simmons, 1996). Evidence suggests, however, that the types of changes recently detected are beyond typical, and researchers now provide evidence for systematic changes in the global environment (IPCC, 1995a).

As mentioned in Chapter Two, recent years have been among the warmest since the beginning of instrumental recordkeeping, with evidence suggesting that these changes are associated with changes in the concentrations of the greenhouse gases (IPCC, 1995a).[1] Greenhouse gases absorb solar radiation, which warms the Earth's atmosphere and increases surface temperature. Carbon dioxide levels, for instance, have steadily increased over the past 40 years, resulting in atmospheric concentrations nearly 30 percent greater than in preindustrial times (Meyer, 1996; see Figure 6.1). The increase in CO_2 concentration heightens the transmission of incoming solar radiation, while lessening its radiation back to space (Simmons, 1996). The results: the "greenhouse" effect and climate change.

To what extent can climate change be attributed specifically to demographic factors? Evidence continues to accumulate suggesting that much of the change in atmospheric gas concentrations is human-induced (e.g., Kaufmann and Stearn, 1997; Tett et al., 1996), although demographic factors influence this aspect of the environment through many routes. Here we mention three. First, human-induced contributions to CO_2 emissions stem from fossil fuel use through energy consumption and industrial production. Since the beginning of the Industrial Revolution, humans have been increasing the release of CO_2 into the atmosphere through the burning of such

---

[1]The primary greenhouse gases are carbon dioxide, methane, nitrous oxide, ozone, CFCs, and water vapor.

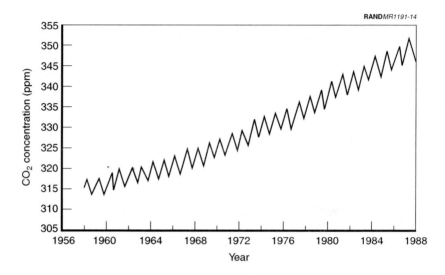

SOURCE: Meyer, 1996, p. 197.

Figure 6.1—Concentration of Atmospheric Carbon Dioxide (in parts per million) Measured at Mauna Loa, Hawaii, 1957–1988

fossil fuels as coal and oil (Goudie and Viles, 1997; IPCC, 1995b). Second, land-use changes, such as deforestation, also affect the exchange of carbon dioxide between the Earth and the atmosphere,[2] in addition to affecting solar radiation, and therefore atmospheric energy balance (Goudie and Viles, 1997). And third, other consumption-related processes, such as paddy rice cultivation and livestock production, are responsible for greenhouse gas releases to the atmosphere, particularly methane releases (Heilig, 1994a).[3]

Although greater numbers of people can mean higher levels of fossil fuel use and more land-use changes, this is an overly simplistic way of viewing the population–climate change relationship. The three

---

[2]To contrast emissions from fossil fuel combustion and deforestation, evidence suggests that tropical deforestation may account for the release of 0.5–4.2 x 10^{12} kilograms per year of carbon to the atmosphere, compared with 5.2 x 10^{12} kilograms per year from fossil fuels (Simmons, 1996, p. 355)

[3]Methane levels have increased from 0.34 parts per million by volume (ppmv) some 160,000 years ago to 1.4 ppmv in 1955 and 1.7 ppmv in 1988 (Heilig, 1994a).

routes of human climatic influence mentioned above suggest that human impacts are related to many factors, including technological change, types of energy use, and even dietary preferences. How then can we understand the relative contribution of population to climate change vis-à-vis these mediating factors? Here a theoretical framework, "IPAT," introduced by Paul Ehrlich and colleagues becomes useful.

The I = P x A x T framework hypothesizes that environmental impact (I) is determined by the interacting effects of population size (P), per capita consumption levels (A, for affluence), and finally the per capita pollution generated by the technology (T) used to satisfy the consumption levels.[4] The equation has been used by several researchers to examine the relative importance of population size and growth in climate change, as compared to other factors (e.g., Bongaarts, 1992; Dietz and Rosa, 1997; Gaffin and O'Neill, 1997). The decomposition exercises tend to reveal that population size and growth are important in the emission of greenhouse gases. One study concludes that population size and growth will account for 35 percent of the global increase in CO_2 emissions anticipated between 1985 and 2100, and 48 percent of the increase in CO_2 emissions from developing nations during that period (see Table 6.1). However, as population growth slows during the next century, its contribution to increases in CO_2 emissions declines, meaning that rising CO_2 emissions will increasingly stem from other factors, such as technological change. Particularly striking is the projected decline in population-driven emissions from developing nations: population growth is estimated to contribute 42 percent of the increase in CO_2 emissions between 1985 and 2020, yet only 3 percent of the increase between 2025 and 2100 (Bongaarts, 1992).[5]

---

[4]Although the simplicity of the IPAT formulation has been criticized (e.g., MacKellar et al., 1998; O'Neill, 2000; Thomas, 1992), it is nonetheless a useful tool in demonstrating the need to consider several factors when examining environmental change. Of particular importance is the idea that each of the equation's components interact in a multiplicative fashion to determine overall environmental impact, as opposed to each representing an independent, additive component.

[5]Bongaarts (1992, p. 308) estimates the contribution of population growth as the proportional reduction in average annual CO_2 emissions that would occur if population size were to remain constant after 1985 and if the projected future trend in the per capita emission rate remained unaffected.

Additional work aiming to quantify the role of population in global climate change has been based on simulation models that examine contrasts between the impact on greenhouse gas emissions of more rapid fertility reduction compared with the impact of lower per capita greenhouse gas emissions. The models indicate that in the near to medium term (to 2050), policies aiming to reduce per capita emissions have a far greater effect than population-oriented policies, although in the longer term (to 2100) slowing fertility can make a substantial contribution to slowing greenhouse gas emissions.[6] Declines in fertility to below replacement level in both more-developed and less-developed countries could reduce projected emissions in 2100 by 37 percent (see Figure 6.2; O'Neill, 2000).

Table 6.1

Projected increases in CO_2 Emissions, and the Estimated Contribution of Population Growth to This Increase, Between 1985 and 2100

	Developing World	Developed World
Increase in CO_2 emission (petagrams (10^{15} grams) of C per year)		
1985–2025	4.7	1.7
2025–2100	7.3	6.4
1985–2100	12.0	8.1
Contribution of population growth to emission increase (percent)		
1985–2025	53.0	42.0
2025–2100	39.0	3.0
1985–2100	48.0	16.0

SOURCE: Bongaarts, 1992, p. 309.

---

[6]Within the simulation exercise, the population-oriented policy option emphasizes policy designed to bring about a more rapid fertility decline (e.g., policies related to female education, family planning programs, and social and economic development). This scenario assumes the total fertility rate in MDCs falls to 1.3 or 1.4 by 2030–2035. In LDCs, fertility also declines, stabilizing at slightly below replacement level by 2030–2035. In the contrasting scenario, policies which aim to reduce per capita greenhouse gas emissions are emphasized (e.g., substitute technology development), resulting in lower per capita emissions of energy-related greenhouse gases (N_2O, CH_4, CO_2).

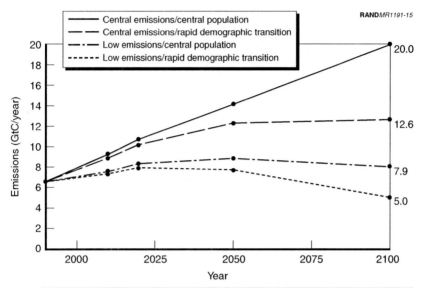

	Population (in millions)				Per capita emissions			
	Central		Rapid		Central		Low	
	LDC	MDC	LDC	MDC	LDC	MDC	LDC	MDC
1990	3,986	1,266	3,986	1,266	0.41	3.92	0.41	3.92
2010	5,623	1,388	5,495	1,355	0.58	4.23	0.46	3.53
2020	6,453	1,428	6,178	1,370	0.68	4.37	0.51	3.48
2050	8,416	1,457	7,200	1,289	0.90	4.52	0.55	2.87
2100	8,942	1,411	5,608	899	1.36	5.57	0.55	2.11

SOURCE: O'Neill, MacKellar, and Lutz, 2000.

Figure 6.2—World Greenhouse Gas Emissions from Commercial Energy, 1990–2100, Under Different Demographic and Emission Scenarios

The importance of technology considerations in efforts to reduce climate change is further illustrated by a set of simulation models focused on energy production and CO_2 emissions in developing nations (Bernstein et al., 1999). These models suggest that CO_2 emissions from power generation in developing countries could triple between 1995 and 2020, if new investments in power generation are made without corresponding improvements in efficiency. Low-emission technologies could reduce the emissions increases by half (Bernstein et al., 1999).

With the aim of ascertaining the feasibility of population-oriented policy responses to climate change, Birdsall (1994) examined the potential climatic impact of reductions in population growth rates. She concluded that the costs of reducing carbon emissions by spending to reduce births compares favorably with the costs of such alternative policies as a carbon tax. The marginal cost of reducing carbon emissions by 10 percent through a carbon tax would be $20 per ton, while the reduction of carbon emissions by lowering population growth through family planning would cost between $4 and $11 per ton. Even less costly, at between $3 and $9 per ton of emission reduction, would be declines in population growth resulting from female education (Birdsall, 1994).[7] In line with these results, some argue that although population-oriented policy may not be the most effective policy for addressing climate change, it would represent a logical component of the "portfolio" of appropriate mitigation efforts (O'Neill et al., 2000).

Population size does not produce climate change alone. Income is also a critical determinant. At the national level, emissions rise as incomes move upward toward a maximum of about $10,000 in per capita GDP, above which emissions decline likely because of improvements in energy efficiencies (Dietz and Rosa, 1997). This relationship demonstrates the U-shaped curve indicative of some aspects of the connection between population and the environment as economies move through the development continuum. The resulting disparities in contribution to climate change are demonstrated by the striking variation in CO_2 emissions across developed and developing nations (see Figure 6.3).

Climate change also provides an excellent example of the role of organizational and institutional response in defining the relationship between population and the environment—and its international aspect highlights the difficulties of development and implementation of cross-national environmental regulation. The atmosphere can be considered a resource predominantly shared by the global community, and, in this sense, it is a common property resource or a

---

[7]Birdsall's analysis does not account for changes in the marginal cost of emission reduction across time. The emission reductions stemming from family planning would accrue over time, while those due to carbon taxes would be more immediate.

"commons." The argument has been made that common property resources are susceptible to overuse because the benefits of resource consumption accrue at the individual level, while the externalities, or costs, are borne by the larger community (Hardin, 1968). As in the

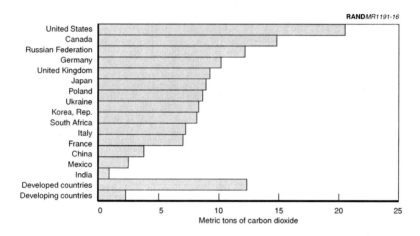

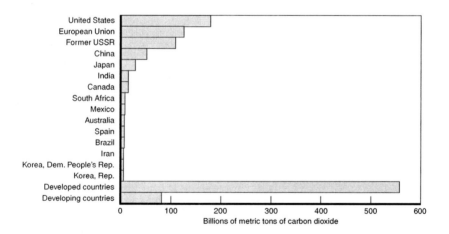

SOURCE: WRI, 1998, p. 176.

Figure 6.3—Per Capita CO_2 Emissions for the 15 Countries with the Highest Total Industrial Emissions, 1995 (top), and Cumulative Carbon Dioxide Emissions, 1950–1995 (above)

case of many common resources, the free market fails as an efficient regulating mechanism because of the lack of congruence between costs and benefits. Specific to climate change, the impacts of global warming are dispersed across the globe, while the benefits of specific emissions (e.g., transportation access, industrial production) accrue to the individual polluters. The issue of management of common property resources and of imposing costs for those who use them remains central to international environmental negotiations today.[8] For climate change, the internalization of costs could occur through taxes on emissions, the development of an emission permit system, or reconsideration of energy subsidies that encourage the development and use of environmentally damaging technologies (IPCC, 1995b).[9]

Consequences of Climate Change

Climate changes feed back into social and ecological systems, affecting both demographic and natural processes. It must be noted, however, that much of the research demonstrating potential effects of climate change is speculative in nature, based on projecting existing knowledge into an unknown future.

Sea Level Rise. Climate change is anticipated to melt glaciers and polar ice and bring warmer ocean temperatures, as well as melting glaciers and polar ice, resulting in a possible sea level rise of 50 centimeters by 2100 (IPCC, 1995a). National land losses are projected to range from 0.05 percent in Uruguay, 1 percent for Egypt, 6 percent for the Netherlands, and up to 17.5 percent for Bangladesh (see Figure 6.4; also see Begum, 1996).

---

[8]The priority owed to this consideration is demonstrated by its inclusion as Principle 16 in *Agenda 21: Programme of Action for Sustainable Development* (United Nations, 1992). The agenda represents the final text of agreements negotiated by governments at the United Nations Conference on Environment and Development in Rio de Janeiro, Brazil, 1992.

[9]The difficulties inherent in dealing with the externalities of emissions became especially apparent during the United Nations Framework Convention on Climate Change held in Kyoto, Japan, during late 1997. Although a last-minute deal was reached in an attempt to control global warming through reductions in greenhouse gas emissions, many criticized the treaty for not going far enough. Additional negotiations remain underway in both international and national settings.

Impacts are also anticipated in coastal areas of California. Shoreline loss may lead to population relocation, bringing intensified demographic, infrastructure, and environmental pressures especially in adjacent coastal areas. Using the Oxnard Plain of Ventura County as an example, Constable et al. (1997) demonstrate that even a minimum case sea level rise would affect commercial piers, marinas, harbors, and fairgrounds, as well as residential areas, a military base, and a game reserve. Indicative of the demographic scale of impacts, the population of the 29 California coastal counties potentially affected by sea-level rise is projected to increase from 26 million to 63 million between 1990 and 2040 (Constable et al., 1997).

Ecological Impacts. As individual species respond to changes in climate, the composition and geographic distribution of such ecosystems as deserts, forests, and wetlands will shift (IPCC, 1995b).

RAND*MR1191-17*

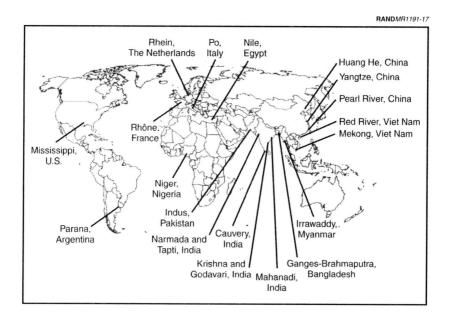

SOURCES: WHO, 1996, p. 155; WRI, 1998, p. 69.

Figure 6.4—Heavily Populated Delta Regions Vulnerable to Sea Level Rise

In turn, agriculture and forestry patterns will likely change. Simulation models examining the potential impact of climate change on world food supplies suggest a 5 percent reduction in global cereal production, assuming a modest level of farm adaptation, such as shifts in planting dates and changes in crop varieties (Rosenzweig and Parry, 1996).[10] While increased productivity is anticipated in some areas, diminished capacity will characterize others; in general, shifts in agricultural productivity would increase the disparities between developed and developing regions (Rosenzweig and Parry, 1996).

Human Health Impacts. Changes in weather patterns, food availability, and water supplies have obvious health implications. In addition, ecosystem shifts change the geography of vector-borne infectious diseases by modifying the location of conditions that exacerbate health risks (see Table 6.2)—the ranges of many diseases will increase as temperatures rise and precipitation patterns change. The most significant changes in disease distribution will occur where diseases are introduced at the edges of historical distributions, areas where residents have little natural immunity. In Africa, for instance, disease distributions may increase to higher elevations formerly too cold to provide habitat for mosquitoes (WRI, 1998). In all, the number of people in the developing world at risk of contracting malaria because of climate change may increase by 5 to 15 percent by 2050 (an increase of 720 million people, representing additional risk above and beyond that anticipated due to population growth) (Martens, 1998).[11]

Population size accounts for a substantial share of the changes in atmospheric gas concentrations because human production and consumption processes represent the forces bringing about such

---

[10]The simulation models assume continuation of trends in economic development and population growth and climate-change scenarios based on the results of three general circulation models with doubled atmospheric carbon dioxide (Rosenzweig and Parry, 1996).

[11]We must keep in mind, however, that quantifying the extent of projected impacts is difficult: data are scarce, assumptions are many, and important interacting factors, such as nutritional status, population density, and health care availability must be taken into account (Martens, 1998).

Table 6.2

Major Tropical Vector-Borne Diseases and the Likelihood of Change in Their Distribution as a Result of Climate Change

Disease	Vector	Number at Risk[a]	Number Infected or New Cases per Year	Present Distribution	Distribution Change?
Malaria	Mosquito	2,400	300 million to 500 million	Tropics/sub-tropics	Highly likely
Schisto-somiasis	Water snail	600	200 million	Tropics/sub-tropics	Very likely
Lymphatic filariasis	Mosquito	1,094	117 million	Tropics/sub-tropics	Likely
African trypto-somiasis	Tsetse fly	55	250,000 to 300,000 cases per year	Tropical Africa	Likely
Dracuncu-liasis	Crusta-cean (copepod)	100	100,000 per year	South Asia, Middle East, Central Africa and West Africa	Unknown
Leishma-niasis	Phleboto-mine sandfly	350	12 million infected, 500,000 new cases per year[b]	Asia, Southern Europe, Africa, and the Americas	Likely
Onchocer-ciasis	Blackfly	123	17.5 million	Africa and Latin America	Very likely
American trypto-somiasis	Triatomine bug	100	18 million to 20 million	Central and South America	Likely
Dengue fever	Mosquito	2,500	50 million per year	Tropics/sub-tropics	Very likely
Yellow fever	Mosquito	450	<5,000 cases per year	Tropical South America and Africa	Very likely

[a]In millions. Top three entries are population-prorated projections based on 1989 estimates.

[b]Annual incidence of visceral leishmaniasis; annual incidence of cutaneous leishmaniasis is 1 million to 1.5 million cases per year.

SOURCE: WRI, 1998, p. 71.

change. One study estimates that population growth will account for 48 percent of the increase in CO_2 emissions anticipated between 1985 and 2100 from developing nations (Bongaarts, 1992). Income levels are also a critical determinant of climate change, with today's developed nations contributing a significantly greater amount of greenhouse gas emissions stemming from energy-intensive consumption patterns.

POPULATION AND LAND-USE CHANGE

Fulfilling the resource requirements of a growing population ultimately requires some form of land-use change, be it to provide for the expansion of food production through forest clearing, to intensify production on already cultivated land, or to develop the infrastructure necessary to support increasing human numbers. Indeed, it is the ability of the human race to manipulate the landscape that has allowed for the rapid pace of contemporary population growth. Agriculture and deforestation are two prominent forms of human-induced land-use change; each is reviewed below.

Agriculture

Agricultural transformation is the human-induced land-use change that has most modified the Earth's surface. Agricultural extensification, meaning increases in cultivated land area, is driven by increasing food demands resultant of a growing human population, in addition to changes in living standards and diets (Bender and Smith, 1997). During the past three centuries, cultivated land area has expanded by more than 450 percent, increasing from 2.65 million square kilometers to 15 million square kilometers. Today the acreage is nearly the size of South America (Meyer, 1996). As agriculture expands, increasingly marginal lands are typically subject to extensification, as farmers move into areas with poor soil, inadequate rainfall, or steep slopes (Shapiro, 1995). In the world's most densely populated nations (e.g., India, China, Indonesia), the cultivated land area is already nearing a maximum, suggesting that future food needs within these areas must be met by means other than extensification (Bongaarts, 1996). An alternative to agricultural extensification is intensification, referring to various means of increasing yields. Examples of intensification include the introduction of irrigation, shortening of fallow periods, and the use of such technological inputs as fertilizers, pesticides, and herbicides.[12, 13]

Although population size is an important determinant of food demand, the type of agricultural land-use change and the implications of that change will be determined by fundamental social, economic, and technological factors (Heilig, 1994b). As a result, the relationship between demographic processes and agricultural land-use change will vary across contexts and will change over time. The different relationships between demographic factors and agricultural land-use change are evidenced by the following examples:

- In Rwanda, population densities are already high, and increasing levels of population pressures have led to land degradation. A high proportion of the Rwandan population is engaged in traditional, subsistence agriculture, and increasing local densities have resulted in the conversion of marginal lands, such as steep hillsides, to agricultural use. To meet heightened food demands, fallow periods are shortened (May, 1995), depleting soil nutrient cycling capacity (Simmons, 1996). Similar conclusions have been reached in Zaire, where research suggests that reductions in population growth would alleviate some of the pressure on the nation's agricultural system (Shapiro, 1995).

- In Honduras, research suggests that environmental destruction is caused more by inequality of resource distribution and patterns of economic development than by population pressures per se. The boom and bust cycles of cotton and cattle industries, largely in response to international markets, have resulted in severe inequalities in access to land. Most small producers are concentrated on steep mountain slopes of marginal agricultural use, while landless peasants rent land in exchange for the labor required to convert additional land to pasture. Here, it is not

---

[12]The total area of cultivated land worldwide increased 466 percent from 1700 to 1980 (Meyer and Turner, 1992, from Matson et al., 1997), and, while expansion has declined, yields continue to increase due to agricultural intensification such as technological developments, although the ability of such inputs to increase yields appears to be waning (WRI, 1998).

[13]Another response to food shortages, unrelated to land-use change in the local context, is increased reliance on food importation, thereby increasing dependence on other economies.

demographic pressure, but the necessity of farming on marginal lands due to inequitable land distribution, that results in agriculturally related land degradation (DeWalt, Stonich, and Hamilton, 1993).

- In Nepal, substantial environmental degradation has occurred from deforestation caused by the expansion of agricultural land use and fuelwood harvest. However, some researchers argue that in specific regions, population growth has provided the impetus for improved land management—in Jhapa, Chitawan, and Morang districts, crop production has outpaced population growth (Subedi, 1997).

In sum, agricultural land-use change can take the form of either extensification or intensification. Case studies demonstrate that the form of change undertaken in a particular locale is influenced by a combination of factors, including population pressures, land tenure policies, and international markets.

Deforestation

Deforestation is linked to agricultural land-use change, as it often is a consequence of agriculture extensification (Houghton, 1994). According to data published by the Food and Agricultural Organization, in 1995 the area of the world's forests was estimated at about 3.5 billion hectares.[14]   A net decline in forest cover of 180 million hectares occurred during the 15-year interval 1980–1995, although changes in forest cover vary greatly across regions. Developed regions experienced a net increase of 20 million hectares from 1980 to 1995; Europe experienced a 4 percent gain in forested area during this time (see Figure 6.5).[15] Yet, a net loss of 200 million hectares occurred in developing regions during 1980–1995. Africa lost 10 percent of its forest cover (FAO, 1998). As in the case of agriculture, patterns of deforestation are the product of complex interactions among demographic factors, policy, and international market pressures.

---

[14]One hectare is equivalent to 2.471 acres.

[15]The increase in forested area noted within developed nations is due primarily to reforestation and natural regrowth on abandoned agricultural land (FAO, 1998).

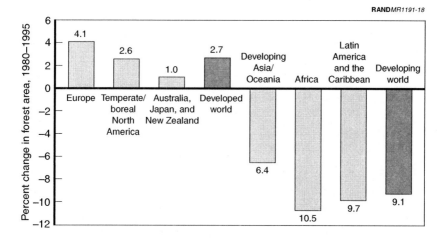

NOTE: Data exclude the countries of the former Soviet Union.
SOURCE: FAO, 1999.

Figure 6.5—Forest Area in 1995 Compared with 1980

Although the "frontiers" of most developed nations have long been closed, frontier expansion continues in many areas of the developing world. As migrants push the boundaries, land may be converted to agriculture or ranching, often through deforestation. For example, the Costa Rican landscape has been radically changed by both deforestation and population growth, especially since World War II. During the postwar era, the nation experienced a fourfold increase in population from 800,000 to more than 3 million inhabitants and about 50 percent of the landscape was cleared of its primary forest cover (see Figure 6.6; Rosero-Bixby and Palloni, 1998; also see Skole et al., 1994, for deforestation estimates).[16] A strong positive correlation exists between the number of nearby potential cultivators and the probability of deforestation—more proximate cultivators means a greater probability of forest loss. The complete relationship between demographic pressure and deforestation is not, however, so

---

[16]It should be noted that protected areas constitute nearly 25 percent of Costa Rican land area.

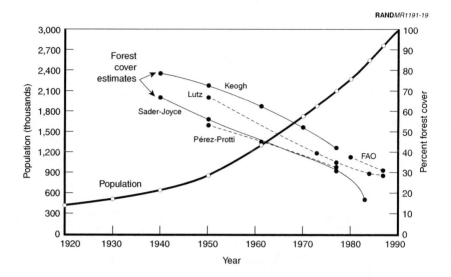

SOURCES: Rosero-Bixby and Palloni, 1998; FAO, 19990; Keogh, 1984; Lutz et al., 1993, pp. 325–331; Pérez and Protti, 1978; Sader and Joyce, 1988, pp. 11–19.

**Figure 6.6—Population and Forest Cover Estimates in
Costa Rica, 1920–1990**

simple. Human pressure to deplete forest cover appears significantly stronger in the case of landless peasants, suggesting that land tenure systems and poverty are important factors in Costa Rican deforestation. Additionally, many other factors suggested as important remain untested. For example, demands from international markets for export crops such as bananas and beef result in land pressure and, ultimately, in loss of forest cover (Rosero-Bixby and Palloni, 1998).

The following case studies provide additional examples of the relationship between demographic factors, mediating forces, and deforestation.

- In Indonesia, the government has been active in frontier settlement and the resulting deforestation. Between 1905 and 1989 nearly 5 million people were moved from densely populated Java to largely forested outer islands. The transmigration projects were designed to relieve Java's population pressure, although the environmental costs have run high—World Bank estimates suggest that transmigration policies have resulted in the deforestation of 750,000 hectares per year, although other estimates reach as high as 1.8 million hectares per year (Fearnside, 1997).[17]

- A historical study of deforestation in Madagascar reveals that, although population growth and shifting cultivation are often held responsible for deforestation, rapid deforestation actually took place at a time when population growth was low and shifting cultivation was banned. Forest clearing was driven by the state's economic objectives, particularly the introduction of coffee as a cash crop (Jarosz, 1993).

- In South America, a mixture of policy, poverty, and demographic pressures has been found responsible for much of the deforestation of the Brazilian Amazon. Driven by government incentives for ranching projects during 1960–1980, conversion of forest lands for pasture was the primary cause of deforestation. However, pressure from commercial lumbering, driven by an increase in global and domestic demand for tropical hardwoods, has now become the dominant force in deforestation. Mining and fossil fuel extraction projects are also under way (Moran, 1992).

As suggested by these examples, many different forces are responsible for changes in land use. Nonetheless, researchers have distinguished four phases of land-use responses in the face of local population growth in the context of developing nations (Bilsborrow and Ogendo, 1992). First, adjustments in tenure arrangements may be made, including fragmentation or redistribution of land parcels. Second, additional land may be appropriated through land extensification, particularly in frontier areas. Third, new technologies may be adopted to increase land productivity. And finally, a demographic response in the form of fertility reduction or out-migration may take place.

---

[17]The estimates vary due to lack of objective verification from satellite imagery and the use of different definitions of deforestation. For example, the definitions of deforestation differ in the degree of logging disturbance necessary to qualify as deforestation, in addition to their treatment of the reclearing of secondary succession by shifting cultivation (Fearnside, 1997).

Given the many influences acting on land-use change and the different responses to demographic pressures, how do we disentangle the specific effect of population? In other words, to what extent can land-use change be attributed specifically to demographic factors? The case studies reviewed above suggest that both population size and distribution play a role in land-use change. In general, countries with higher population densities have greater proportions of cultivated land area and associated deforestation (Bilsborrow, 1992; Bilsborrow and Ogendo, 1992). Research also suggests that local population growth may reduce farm size, thereby threatening the ability of rural farm families to derive subsistence from their holdings. These issues are particularly critical in areas already experiencing significant inequalities in access to land (Bilsborrow and Stupp, 1997; Muhammed, 1996). Finally, increases in local population densities can lead to changes in agricultural production techniques, which may increase food production (Subedi, 1997) but also bring concerns with eventual environmental consequences (Shapiro, 1995).

Specifically related to deforestation, a cross-national study of the relationship between population and deforestation found population size and growth related to national rates of forest loss, especially in Burundi, Rwanda, and Haiti, which have small rain forests. In countries with large forests, such as Brazil, capital investments also appear associated with higher levels of deforestation (Rudel, 1989).

Although population size, growth, and distribution are the most important demographic forces in land-use change, other demographic factors may also influence land conversion, particularly as related to food demand. Basic caloric requirements vary according to an individual's age and gender, with daily demands ranging from less than 1,000 calories per day for infants to about 3,900 for an adult man under extreme physical stress (Bender and Smith, 1997). Illustrating a specific regional example of demographic effects on food demand, Table 6.3 shows FAO estimates of the effects of demographic change on food requirements. These data imply, for example, that the change in age structure will account for 7 percentage points of the expected 214 percent increase in food energy requirements in Africa between 1995 and 2050. Urbanization, too, can be related to food demand. Rural dwellers tend to be more physically active than urban dwellers, meaning that the shift of population from rural to urban areas will lower regional caloric needs. This effect is expected to be modest, however. For instance, between 1995 and 2050, food demand in Asia is expected to increase 69 percent. This represents four percentage points below the increase that would be anticipated if no further urbanization occurred (Bender and Smith, 1997). In general, Table 6.3 provides evidence of the overwhelming role of population size in determining changes in future food demand. In each region presented, population growth is projected to account for at least 90 percent of the overall change in food requirements, 1995–2050.

Finally, people demand larger quantities of food and more dietary variety as incomes rise (Heilig, 1995). Of particular environmental consequence, global meat consumption is expected to expand by 75 percent between 1990 and 2020, attributable to rising incomes and changing dietary preferences (Bender and Smith, 1997). Dietary changes are already noticeable—in Asia, where many countries have

Table 6.3

Effect of Demographic Change on Food Requirements, by Region, 1995–2050

Type of Change	Africa	Asia	Europe	Latin America/ Carib- bean	North America	Oceania
	Percentage Change in Food Requirements, 1995–2050					
All demographic effects	+214	+69	–9	+80	+31	+61
Population growth	+194	+66	–7	+74	+33	+61
Combined effects[a]	+7	+2	–2	+3	–1	0
Older age structure	+7	+2	–1	+2	0	0
Increased height	+2	+2	0	+2	0	+1
Smaller % of women pregnant	0	–1	0	0	0	0
Greater % urban	–3	–4	–1	–2	–1	–1

[a]Includes factors not listed separately.
SOURCES: Bender and Smith, 1997; FAO and UNFPA, 1996.

a vegetarian tradition, the production of slaughtered meat increased by 470 percent between 1961–1963 and 1989–1991, while the population grew by 80 percent (FAO, 1994, as presented in Heilig, 1994b). The environmental implications of these dietary shifts arise because livestock are inefficient users of grain—the amount of grain it takes to produce one calorie of beef would yield 10 calories if eaten directly by humans (Bender and Smith, 1997). As such, significantly higher levels of grain production will be required to meet heightened consumer demand for beef, demonstrating an important link between income levels, diet, and changes in agricultural production.

Consequences of Land-Use Change

Land-use change in the form of agriculture and deforestation has several ecological impacts. Agriculture can lead to soil erosion, while overuse of chemical inputs can also degrade soil (Matson et al., 1997). Deforestation also increases soil erosion, lessening the ability of soils to hold water and increasing the frequency and severity of floods (Goudie and Viles, 1997). Land-use change in general results in habitat loss and fragmentation, the primary cause of contemporary species decline (Wilson, 1992). If current rates of forest clearing continue, it has been estimated that one-quarter of all species on Earth could be lost within the next 50 years (Simmons, 1996).

As in the case of climate change, the quantification of the exact role of population in land-use change is difficult, although humans have affected nearly all of the Earth's land surfaces. Population pressure in the form of high densities appears related to agricultural expansion and deforestation; yet how that pressure is manifested will be influenced by social, economic, and political processes. Context matters. Whether an individual farmer chooses to clear additional land or intensify cultivation will be greatly influenced by government policies regarding land-use and ownership, particularly as they relate to the distribution of wealth and resources within the nation. And in some cases, although humans are agents of land-use change, they are driven not by subsistence but by the forces of economic development and the international market.

Chapter Seven

CONCLUSIONS AND IMPLICATIONS

The following concluding comments are divided into three categories: general policy concerns, specific policy implications, and future research needs.

GENERAL POLICY CONSIDERATIONS

Environmental policies should consider both demographic concerns and mediating factors. Informed policymaking on environmental issues requires sophisticated approaches that recognize the important interactions among demographic, environmental, and mediating factors, while also remaining sensitive to context. For example, reductions in population growth would bring cost-effective, long-term benefits to the global climate (Birdsall, 1994). Yet population is not the only important consideration; consumption patterns and economic development also play key roles in determining greenhouse gas emissions (Dietz and Rosa, 1997). As a result, population reduction, consumer education, and incentives aimed at the development of alternative energy sources could all contribute to future emission reduction.

International efforts will be required. Ecosystems do not abide by national boundaries. As a result, many environmental issues are inherently global and will require international commitments and cooperation. We can look to the Montreal Protocol of 1987 as an example of the potential of such agreements, in addition to the difficulties inherent in pursuing them.

The role of international markets in environmental degradation should be recognized. Policy development should bear in mind that many localized environmental changes are caused by pressures beyond the proximate context. For example, international markets for cash crops, such as coffee, acted as an important factor in historical rates of deforestation in Madagascar (Jarosz, 1993).

Relevant policies can occur at many levels. International policies are certainly not the only relevant venue for environmental negotiation. Many local and national factors act to mediate the relationship between population and the environment. As such, site-specific issues must also be addressed.

SPECIFIC POLICY IMPLICATIONS

The negative environmental implications of global population size suggest the environmental relevance of slowing population growth. As for population distribution, careful planning should accompany shifts in population distribution, particularly migration flows to ecologically sensitive areas. In addition, lessening the "push" factors that act on migrants may reduce motivations to move.

With regard to compositional factors, although age composition is not amenable to policy intervention per se, nonpopulation-oriented policy response could be relevant for the potential environmental implications of age composition (e.g., retirement migration pressures reduced through land-use restrictions). Policy should specifically address environmentally damaging patterns of production and consumption as they relate to population composition.

More specific policy implications include the following:

Family planning policies that enable couples to avoid unwanted pregnancies would reduce fertility and rates of population growth (Bulatao, 1998), therefore reducing pressure on environmental resources. This would be particularly beneficial in areas already characterized by resource scarcity. For example, in developing regions, nearly 75 percent of the increase in fuelwood demand from 1980–2000 was roughly estimated to be due to local population growth. Reductions in fertility rates would slow population growth and ease local environmental burdens, such as fuelwood shortages.

Rural development policies could reduce rural-to-urban migration, perhaps easing pressure on urban infrastructures. This would be particularly beneficial in areas where rural resource shortages or lack of rural opportunities act as "push" factors fueling rapid urban growth. Nearly 50 percent of the world's 6 billion people live in urban areas, and, by 2030, it is expected that nearly 5 billion (61 percent) of the world's people will live in cities. About half of this massive urban transition is fueled by migration, a product of both "pull" forces, such as urban opportunity, and "push" factors, such as rural resource scarcity.

More equitable land-tenure policies could ease resource pressures and, therefore, reduce agricultural extensification and rural-to-urban migration. This would be especially beneficial in areas characterized by subsistence agriculture where individuals lack access to land. In Honduras, for example, severe inequalities in access to land result in the use of steep mountain slopes for agricultural production (DeWalt, Stonich, and Hamilton, 1993). More equitable land access would lessen pressure to cultivate marginally productive areas.

Policies encouraging sustainable intensification of land resources could increase yields, thereby lessening the need for agricultural extensification. This would be especially beneficial in areas characterized by arable land shortages. Technological innovation, or the intensified use of existing technologies, is a recognized phase in land-use change resulting from demographic pressure (see Bilsborrow and Ogendo, 1992). In northern Nigeria, for instance, where population densities are high and the majority of land is already under cultivation, improved land management has increased agricultural yields (Mortimore, 1993). Encouragement of sustainable intensification would assist in meeting long-term food production needs.

Careful planning must accompany change in local population densities. This is true both in rapidly growing megacities, as well as in less densely populated areas receiving large influxes of migrants. Coastal regions of the United States, for instance, are at risk of ecological decline because of the recent increases in amenity-driven migration. Between 1980 and 1990, North Carolina's narrow coastal islands of Bodie and Hatteras absorbed most of Dare County's 280 percent growth (Bartlett, Mageean, and O'Connor, 2000). Careful planning is required to ameliorate the environmental decline potentially brought on by these shifts in population distribution.

Policies providing incentives for the development of sustainable production processes could ease environmental pressure. Both developed and developing regions could benefit from the application of improved technological efficiencies. Environmental pressures vary in important ways across the continuum of economic development. Early development stages are typically characterized by increasing environmental pressures as a result of the adoption of energy-intense production and consumption patterns and the lack of pollution control regulatory mechanisms. As economies move along the development continuum, however, pollution controls tend to be implemented, and efficiencies in technology are gained. Technological advancements could greatly reduce the environmental implications of economic development in today's developing regions, while also providing benefit to today's more-developed economies. In the long run, the development of low-emission technologies may provide economic benefit as well (Bernstein, 1999).

Policies providing education and encouragement for sustainable consumption could ease environmental pressure. This is particularly true in areas where consumption processes are environmentally intensive. The excessive fossil fuel dependence, for instance, of the more-developed nations, in combination with the growing energy demands of the less-developed world, mandate the development, implementation, and encouraged use of sustainable forms of energy.

RESEARCH NEEDS

Several insightful overviews have recently been written by both social and natural scientists on the topic of population and the environment (see appendix). In nearly every case, the authors criticize the academic community for the lack of research attention paid to these important relationships. While natural scientists studying environmental processes have often neglected human dimensions of change, social scientists examining demographic processes have neglected the importance of environmental context. Our quest for a better understanding of the relationship between population and the environment must, therefore, be expanded, especially with research along interdisciplinary lines. Teams of natural and social science researchers must be supported to cooperate in unraveling the myriad processes leading to various changes in environmental context. Specific suggestions include the following:

Encourage interdisciplinary research. A more precise scientific understanding of the complex interaction between demographic processes and the environment is needed. To accomplish this goal, natural and social scientists must work together to make the best use of contemporary technology. This integration will continue the encouraging movement toward a truly interdisciplinary global environmental science. The use of such modern technology as remote sensing to study environmental change across time is especially promising (Liverman et al., 1998; see appendix).

Continue to collect data and develop new models that allow links between natural and social processes. Research on population and the environment has continually been hampered by a lack of appropriate data, and significant concerns persist with regard to data availability and quality (Bilsborrow, 1992). Support must be provided for the collection and provision of relevant information allowing development of local, regional, national, and global scale models of the interrelations between natural and social processes.

Work at various levels of analysis to inform relevant policy circles. Research investigating the relationship between population and the environment at various scales of analysis can inform policymakers operating in these different arenas. For instance, micro-level studies of local area interactions between population and the environment provide important insight into local processes and relevant policy. However, macro-scale analyses of global processes provide critical information for use in the international policy context.

Examine the environmental implications of population composition. In particular, a more precise understanding of the links between age composition and issues related to consumption and migration would help policymakers better address the mechanisms that lead to environmental change. Which environmental consequences arise as the young generation in developing nations comes of age? What role does the aging population of more-developed regions play in the environmental future?

Recognize and examine the dynamic nature of environmental processes. Both natural and social systems are constantly in flux; the changes in each bringing reciprocal changes in the other. Recall, for example, the variation in types of environmental pressures that characterize societies at different stages of economic development. Sanitation, water quality, and indoor air pollution are critical environmental issues in less-developed regions, while technologically driven hazards dominate environmental concern within developed contexts. Each of these offers important arenas for exploring the role of demographic factors.

Carefully consider intervening factors. It is important, again, not to consider *only* the direct link between demographic factors and environmental conditions but also how those relationships are mediated by technology, institutions, the policy context, and cultural values and norms. One of the primary contributions of demographers to the discussion of human-induced environmental impacts has related to the efficacy of family planning approaches, under the assumption that lower population growth lessens resource pressure (Pebley, 1999). Additional research should be undertaken to ascertain the relevance of policy measures in different arenas—for instance, population distribution or policies targeting production and consumption patterns. As stated by Hogan (1992; see appendix), consideration must go beyond population growth as "villain" to examine reciprocal effects and other factors.

Critically examine the ability of the free market to regulate the use of common property resources. The costs and benefits of the use of "commons" must be carefully considered. Whenever possible, the polluter should bear the cost of pollution, thereby having an incentive *not* to pollute. The atmosphere, for instance, is a global commons and the costs of using the atmosphere as a sink should be internalized. This could occur, for example, through the development of an emission permit system in which companies able to reduce CO_2 output beyond the set target can sell their remaining emission credits, thereby providing an economic incentive to reduce pollution. Such a proposal is outlined in the Kyoto Protocol.

Be mindful of the reciprocal nature of the relationship between population and the environment. As individuals become increasingly aware that the health of the planet is central to the health of its inhabitants, more public and policy attention is being paid to these issues. For instance, heightened public awareness of the links between air pollution and human health increase the urgency for regulatory mechanisms targeting emissions. Let us be mindful of these two-way relationships between demographic factors and the environment as we continue to work toward creating an environmentally sustainable human population.

The environmental implications of demographic dynamics are obviously complicated and can sometimes be controversial. While some view population growth in developing regions as the primary culprit in environmental decline, others focus on the costly environmental effects of consumption among the developed nations. Such differing emphases can lead to a disagreement about the most effective and equitable policy solution—slow population increase in less-developed nations or lessen destructive production and consumption patterns of the more-developed nations? Such a debate, however, presumes that a one-step solution to the complex realities of the relationship between population and the environment exists. As demonstrated by the foregoing discussion, both population growth and consumption play a role in environmental change and are among the many factors that should be considered and incorporated in realistic policy debate and prescriptions. Other demographic factors, such as distribution and age and sex composition, are also of relevance.

In the end, many causes underlie contemporary environmental degradation, and only some are demographic in nature. Yet population does matter, and increased attention to the environmental implications of demographic dynamics can improve policy capacity to respond to contemporary environmental change.

REFERENCES

Acreman, Michael, "Principles of Water Management for People and the Environment," in Alex de Sherbinin and Victoria Dompkin, eds., *Water and Population Dynamics: Case Studies and Policy Implications*, Washington, D.C.: American Association for the Advancement of Science, 1998, pp. 25–48.

Bartlett, Jay G., D. M. Mageean, and R. J. O'Connor, "Residential Expansion as a Continental Threat to U.S. Coastal Ecosystems," *Population and Environment: A Journal of Interdisciplinary Studies*, Vol. 21, No. 5, May 2000, pp. 429–468.

Begum, Selina, "Climate Change and Sea Level Rise: Its Implications in the Coastal Zone of Bangladesh," *Global Change, Local Challenge*, HDP Third Scientific Symposium, September 20–22, 1995, Vol. 2, Poster Papers, Human Dimensions of Global Environmental Change Programme, Geneva, Switzerland, 1996.

Bender, William, and Margaret Smith, "Population, Food, and Nutrition," *Population Bulletin*, Vol. 51, No. 4, 1997.

Benneh, George, "Environmental Consequences of Different Patterns of Urbanization," in *Population, Environment and Development: Proceedings of the United Nations Expert Group Meeting on Population, Environment and Development*, New York: United Nations, 1994.

Bernstein, Mark, Pam Bromley, Jeff Hagen, Scott Hassell, Robert Lempert, Jorge Munoz, and David Robalino, *Developing Countries and Global Climate Change: Electric Power Options for Growth*,

Arlington, Va.: Pew Center on Global Climate Change, 1999, URL: http://www.pewclimate.org/projects/pol_countries.html (retrieved March 2000).

Berry, Brian J. L., "Urbanization," in B. L. Turner II, William C. Clark, Robert W. Kates, John F. Richards, Jessica T. Mathews, William B. Meyer, eds., *The Earth as Transformed by Human Action: Global and Regional Changes in the Biosphere over the Past 300 Years,* Cambridge, Mass.: Cambridge University Press, 1990.

Bilsborrow, Richard E., "Population Growth, Internal Migration, and Environmental Degradation in Rural Areas of Developing Countries," *European Journal of Population,* Vol. 8, 1992, pp. 125–148.

Bilsborrow, Richard E., and Daniel Hogan, *Population Deforestation in the Humid Tropics,* Liege, Belgium: International Union for the Scientific Study of Population, 1999.

Bilsborrow, Richard E., and H. W. O. Okoth Ogendo, "Population-Driven Changes in Land Use in Developing Countries," *Ambio,* Vol. 21, No. 1, 1992, pp. 37–45.

Bilsborrow, Richard E., and Paul Stupp, "Demographic Processes, Land, and the Environment in Guatemala," in Anne R. Pebley and Luis Rosero-Bixby, eds., *Demographic Diversity and Change in the Central American Isthmus,* Santa Monica, Calif.: RAND, 1997, pp. 581–615.

Birdsall, Nancy, "Another Look at Population and Global Warming," in *Population, Environment and Development: Proceedings of the United Nations Expert Group Meeting on Population, Environment and Development,* New York: United Nations, 1994.

Bongaarts, John, "Population Growth and Global Warming," *Population and Development Review,* Vol. 18, No. 2, 1992, pp. 299–319.

_____, "Population Pressure and the Food Supply System in the Developing World," *Population and Development Review,* Vol. 22, No. 3, 1996, pp. 483–503.

Boserup, Ester, *The Conditions of Agricultural Growth: The Economics of Agrarian Change Under Population Pressure,* Chicago, Ill.: Aldine Publishing Company, 1965.

_____, "Environment, Population and Technology in Primitive Societies," *Population and Development Review*, Vol. 1, No. 1, 1976, pp. 21–36.

_____, *Population and Technological Change: A Study of Long-Term Trends*, Chicago, Ill.: The University of Chicago Press, 1981.

Brennan, Ellen M., "Air/Water Pollution Issues in the Mega-Cities," in Shridath Ramphal and Steven W. Sinding, eds., *Population Growth and Environmental Issues*, Westport, Conn.: Praeger Publishers, 1996.

Brown, Lester, *Beyond Malthus: Nineteen Dimensions of the Population Challenge*, New York: W. W. Norton and Company, 1999.

Brown, Lester, Michael Renner, and Brian Halweil, *Vital Signs: 1999*, New York: W. W. Norton and Company, 1999.

Bulatao, Rodolfo A., *The Value of Family Planning Programs in Developing Countries*, *Population Matters* Series, Santa Monica, Calif.: RAND, 1998.

Campbell, Martha M., "Schools of Thought: An Analysis of Interest Groups Influential in International Population Policy," *Population and Environment: A Journal of Interdisciplinary Studies*, Vol., 19, No. 6, 1998, pp. 487–512.

Clarke, John I., "The Impact of Population Change on Environment: An Overview," in Bernardo Colombo, Paul Demeny, and Max F. Perutz, eds., *Resources and Population: Natural, Institutional, and Demographic Dimensions of Development*, Oxford, U.K.: Clarendon Press, 1996, pp. 254–268.

Cohen, Joel E., *How Many People Can the Earth Support?* New York: W. W. Norton and Company, 1995.

Colombo, Umberto, "Energy Resources and Population," in Bernardo Colombo, Paul Demeny, and Max F. Perutz, eds., *Resources and Population: Natural, Institutional, and Demographic Dimensions of Development*, Oxford, U.K.: Clarendon Press, 1996, pp. 53–63.

Constable, A., M. D. Van Arsdol, Jr., D. J. Sherman, J. Wang, P. A. McMullin-Messier, and L. Rollin, "Demographic Responses to Sea Level Rise in California," *World Resources Review*, Vol. 9, No. 1, 1997, pp. 32–44.

Cramer, James C., "Population Growth and Air Quality in California," *Demography*, Vol. 35, No. 1, 1998, pp. 45–56.

Cruz, Maria Concepcion, "Effects of Population Pressure and Poverty on Biodiversity Conservation in the Philippines," in R. K. Pachauri and L. F. Qureshy, eds., *Population, Environment, and Development*, New Delhi: Tata Energy Research Institute, 1997, pp. 69–94.

Cruz, Maria Concepcion, C. A. Meyer, R. Repetto, and R. Woodward, *Population Growth, Poverty, and Environmental Stress: Frontier Migration in the Philippines and Costa Rica*, Washington, D.C.: World Resources Institute, 1992.

Culliton, Thomas J., "Population: Distribution, Density, and Growth," *NOAA's State of the Coast Report*, Silver Spring, Md.: National Oceanic and Atmospheric Administration, URL: http://state-of-coast.noaa.gov/bulletins/html/pop_01/pop.html, 1998 (retrieved March, 1999).

Culliton, Thomas J., Maureen A. Warren, Timothy R. Goodspeed, Davida G. Remer, Carol M. Blackwell, and John J. McDonough III, *50 Years of Population Change Along the Nation's Coasts, 1960–2010*, Rockville, Md.: National Oceanic and Atmospheric Administration, 1990.

DasGupta, Parth S., "Population, Poverty and the Local Environment," *Scientific American*, Vol. 272, No. 2, February 1995, pp. 40–45.

DaVanzo, Julie, and David M. Adamson, *Family Planning in Developing Countries: An Unfinished Success Story*, Santa Monica, Calif.: RAND, Issue Paper, *Population Matters* Series, 1999.

Deininger, Klaus W., and Bart Minten, "Poverty, Policies, and Deforestation: The Case of Mexico," *Economic Development and Cultural Change*, Chicago, Ill: University of Chicago, 1999.

Demeny, Paul, "Population," in B. L. Turner II, William C. Clark, Robert W. Kates, John F. Richards, Jessica T. Mathews, and William B. Meyer, eds., *The Earth as Transformed by Human Action: Global and Regional Changes in the Biosphere over the Past 300 Years*, Cambridge, Mass.: Cambridge University Press, 1990.

De Souza, Roger-Mark, *Household Transportation Use and Urban Air Pollution: A Comparative Analysis of Thailand, Mexico, and the United States*, New York: Population Reference Bureau, 1999.

DeWalt, Billie R., Susan C. Stonich, and Sarah L. Hamilton, "Honduras: Population, Inequality, and Resource Destruction," in Carole L. Jolly and Barbara Boyle Torrey, eds., *Population and Land Use in Developing Countries*, Committee on Population, Commission on Behavioral and Social Sciences and Education, National Research Council, Washington, D.C.: National Academy Press, 1993.

Dietz, Thomas, and Eugene A. Rosa, "Effects of Population and Affluence on CO_2 Emissions," *Proceedings of the National Academy of Science*, Vol. 94, 1997, pp. 175–179.

Dompka, Victoria, ed., *Human Population, Biodiversity and Protected Areas: Science and Policy Issues*, Washington, D.C.: American Association for the Advancement of Science, 1996.

Faber, Carol S., *Geographical Mobility: March 1996 to March 1997 (Update)*, Current Population Reports, Population Characteristics, P20–510, Washington, D.C.: U.S. Bureau of the Census, Issued July 1998.

Falkenmark, Malin, "Rapid Population Growth and Water Scarcity: The Predicament of Tomorrow's Africa," in K. Davis, and M. Bernstam, eds., *Resources, Environment, and Population: Present Knowledge and Future Options*, New York: Oxford University Press, 1991, pp. 81–94.

_____, "Population, Environment, and Development: A Water Perspective," in *Population, Environment and Development*, Proceedings of the United Nations Expert Group Meeting on Population, Environment and Development, United Nations Headquarters, January 20–24, 1992, convened as part of the substantive preparations for the International Conference on Population and Development, 1994, pp. 99–116.

Falkenmark, Malin, and Riga Adiwoso Suprapto, "Population-Landscape Interactions in Development: A Water Perspective to Environmental Sustainability," *Ambio*, Vol. 21, No. 1, February 1992, pp. 31–36.

Falkenmark, Malin, and Carl Widstrand, "Population and Water Resources: A Delicate Balance," *Population Bulletin*, Vol. 47, No. 3, November 1992.

Fearnside, Philip M., "Transmigration in Indonesia: Lessons from Its Environmental and Social Impacts," *Environmental Management*, Vol. 21, No. 4, 1997, pp. 553–570.

Food and Agricultural Organization (FAO), *Forest Resources Assessment, 1990: Tropical Countries*, Rome, Italy: FAO, 1990.

_____, *The Right to Food in Theory and Practice*, Rome, Italy: FAO, 1998.

_____, *The State of the World's Forests, 1999*, Rome, Italy: FAO, 1998.

Fortmann, Louise, Camille Antinori, and Nontokozo Nabane, "Fruits of Their Labors: Gender, Property Rights, and Tree Planting in Two Zimbabwe Villages," *Rural Sociology*, Vol. 62, No. 3, 1997, pp. 295–314.

Frey, William H., "The New Geography of Population Shifts: Trends Toward Balkanization," in Reynolds Farley, ed., *State of the Union: America in the 1990s*, New York: Russell Sage Foundation, 1995, pp. 271–336.

Gaffin, Stuart R., and Brian C. O'Neill, "Population and Global Warming With and Without CO_2 Targets," *Population and Environment: A Journal of Interdisciplinary Studies*, Vol. 18, No. 4, 1997, pp. 389–413.

Glazovsky, Nikita F., "The Aral Sea Basin," in Jeanne X. Kasperson, Roger E. Kasperson, and B. L. Turner II, eds., *Regions at Risk: Comparisons of Threatened Environments*, New York: U.N. Publications, 1995.

Goudie, Andrew, and Heather Viles, *The Earth Transformed: An Introduction to Human Impacts on the Environment,* Oxford, U.K.: Blackwell Publishers, 1997.

Hamilton, Lawrence C., and Carole L. Seyfrit, "Resources and Hopes in Newfoundland," *Society and Natural Resources,* Vol. 7, No. 6, 1994, pp. 561–578.

Hardin, Garrett, "The Tragedy of the Commons," *Science,* Vol. 162, No. 13, 1948, pp. 1243–1248.

Hart, John, *Storm Over Mono: The Mono Lake Battle and the California Water Future,* Berkeley, Calif.: University of California Press, 1996.

Heilig, Gerhard K., "The Greenhouse Gas Methane (CH_4): Sources and Sinks, the Impact of Population Growth, Possible Interventions," *Population and Environment: A Journal of Interdisciplinary Studies,* Vol. 16, No. 2, 1994a, pp. 109–137.

_____, "Neglected Dimensions of Global Land-Use Change: Reflections and Data," *Population and Development Review,* Vol. 20, No. 4, December 1994b, pp. 831–859.

_____, "Lifestyles and Global Land-Use Change: Data and Theses," IIASA WP 95–91, 1995.

Hogan, Daniel Joseph, "The Impact of Population Growth on the Physical Environment," *European Journal of Population,* Vol. 8, 1992, pp. 109–123.

Houghton, R. A., "The Worldwide Extent of Land-Use Change," *Bioscience,* Vol. 44, No. 5, pp. 305–313.

IPCC (International Panel on Climate Change), Summary for Policymakers, The Science of Climate Change—IPCC Working Groups I, II, and III, New York: UNEP, 1995a.

_____, IPCC Second Assessment Synthesis of Scientific-Technical Information Relevant to Interpreting Article 2 of the U.N. Framework Convention on Climate Change, New York: UNEP, 1995b.

Jarosz, Lucy, "Defining and Explaining Tropical Deforestation: Shifting Cultivation and Population Growth in Colonial Madagascar (1896–1940)," *Economic Geography,* Vol. 69, No. 4, 1993, pp. 366–379.

Jasanoff, Sheila, and Brian Wynne, "Science and Decisionmaking," in Steve Rayner and Elizabeth L. Malone, eds., *Human Choice and Climate Change: Resources and Technology,* Vol. 1, Columbus, Ohio: Battelle Press, 1998.

Jolly, Carole L., "Four Theories of Population Change and the Environment," *Population and the Environment: A Journal of Interdisciplinary Studies,* Vol. 16, No. 1, 1994, pp. 61–90.

Kaufmann, Robert, and David Stearn, "Evidence for Human Influence on Climate from Hemispheric Temperature Relations," *Nature,* Vol. 388, 1997, pp. 39–44.

Kellert, Stephen R., "Japanese Perceptions of Wildlife," *Conservation Biology,* Vol. 5, No. 3, September 1991.

_____, "Attitudes, Knowledge, and Behavior Toward Wildlife Among the Industrial Superpowers: United States, Japan, and Germany," *Journal of Social Issues,* Vol. 49, No. 1, 1993, pp. 53–69.

_____, "Social and Perceptual Factors in Endangered Species Management," *Journal of Wildlife Management,* Vol. 49, No. 2, 1985, pp. 528–536.

_____, "Population and Sustainable Development: Distinguishing Fact and Preference Concerning the Future Human Population and Environment," *Population and Environment: A Journal of Interdisciplinary Studies,* Vol. 14, No. 5, 1993, pp. 441–461.

Keogh, R. M., "Changes in the Forest Cover of Costa Rica Throughout History," *Turrialba,* Vol. 34, No. 3, 1984, pp. 325–331.

Liu, Jianguou, Zhiuyun Ouyang, Yingchun Tan, Jian Yang, and Heming Zhang, "Changes in Human Population Structure: Implications for Biodiversity Conservation," *Population and Environment: A Journal of Interdisciplinary Studies,* Vol. 21, No. 1, 1999, pp. 45–58.

Liverman, Diana, Emilio F. Moran, Ronald R. Rindfuss, and Paul C. Stern, eds., *People and Pixels: Linking Remote Sensing and Social*

Science, Committee on the Human Dimensions of Global Change, Commission on Behavioral and Social Sciences and Education, National Research Council, Washington, D.C.: National Academy Press, 1998.

Locher, Uli, "Migration and Environmental Change in Costa Rica Since 1950," in Anne R. Pebley and Luis Rosero-Bixby, eds., *Demographic Diversity and Change in the Central American Isthmus,* Santa Monica, Calif.: RAND, 1997, pp. 667–693.

Long, Larry, *Migration and Residential Mobility in the United States,* New York: Russell Sage Foundation, 1988.

Lutz, Wolfgang, and Einar Holm, "Mauritius: Population and Land Use," in Carole L. Jolly and Barbara Boyle Torrey, eds., *Population and Land Use in Developing Countries,* Committee on Population, Commission on Behavioral and Social Sciences and Education, National Research Council, Washington, D.C.: National Academy Press, 1992.

Lutz, E., M. Vedova, H. Martínez, L. San Román, R. Vázquez, A. Alvarado, L. Merino, R. Celis, and J. Huising, "Interdisciplinary Fact-Finding on Current Deforestation in Costa Rica," environment working paper, Washington, D.C.: The World Bank, 1993.

MacKellar, F. Landis, Wolfgang Lutz, A. J. McMichael, and Astri Suhrkein, "Population and Climate Change," in Steve Rayner and Elizabeth L. Malone, eds., *Human Choice and Climate Change: The Societal Framework,* Vol. 1, Columbus, Ohio: Battelle Press, 1998, pp. 89–193.

MacKellar, F. Landis, W. Lutz, C. Prinz, and A. Goujon, "Population, Households, and CO_2 Emissions," *Population and Development Review,* Vol. 21, No. 4, 1995, pp. 849–865.

Malthus, Thomas Robert, *Population: The First Essay,* 1798, reprint, Ann Arbor, Mich.: University of Michigan Press, 1959.

Martens, Pim, *Health and Climate Change: Modelling the Impacts of Global Warming and Ozone Depletion,* London: Earthscan Publications Ltd., 1998.

Martin, Philip, and Jonas Widgren, "International Migration: A Global Challenge," *Population Bulletin*, Vol. 51, No. 1, 1996.

Matson, P. A., W. J. Parton, A. G. Power, and M. J. Swift, "Agricultural Intensification and Ecosystem Properties," *Science*, Vol. 277, July 25, 1997, pp. 504–509.

May, John, "Policies on Population, Land Use, and Environment in Rwanda," *Population and Environment: A Journal of Interdisciplinary Studies*, Vol. 16, No. 4, 1995, pp. 321–334.

McAtee, Jerry W., and D. Lynn Drawe, "Human Impact on Beach and Foredune Microclimate on North Padre Island, Texas," *Environmental Management*, Vol. 5, No. 2, 1981, pp. 121–134.

McGranahan, David A., *Natural Amenities Drive Rural Population Change*, Washington, D.C.: Food and Rural Economics Division, Economic Research Service, U.S. Department of Agriculture, Agricultural Economic Report No. 781, 1999.

Meadows, Donella H., Dennis L. Meadows, Jorgen Randers, *Beyond the Limits: Confronting Global Collapse, Envisioning a Sustainable Future*, White River Junction, Vt.: Chelsea Green Publishing Company, 1992.

Meyer, William B., *Human Impact on the Earth*, Cambridge, U.K.: Cambridge University Press, 1996.

Meyer, William B., and B. L. Turner II, "Human Population Growth and Global Land-Use/Cover Change," *Annual Review of Ecology and Systematics*, Vol. 23, 1992, pp. 39–61.

Micklin, P. P., "The Desiccation of the Aral Sea: A Water Management Disaster in the Soviet Union," in A. Goudie, ed., *The Human Impact Reader: Readings and Case Studies*, Oxford, U.K.: Blackwell Publishers, 1997.

Mono Lake Committee (MLC), *Political Chronology of the Mono Lake Water Issue*, 1999, URL: http://www.monolake.org/political history/polchr.htm (retrieved February 2000).

Mooney, Linda A., David Knox, and Caroline Schacht, *Understanding Social Problems*, Minneapolis/St. Paul: West Publishing Company, 1996.

Moran, Emilio F., "Deforestation in the Brazilian Amazon," Occasional Paper No. 10, Series on Environment and Development, Bloomington, Ind.: Indiana University, 1992.

Mortimore, M., "Northern Nigeria: Land Transformation Under Agricultural Intensification," in Carole L. Jolly and Barbara Boyle Torrey, eds., *Population and Land Use in Developing Countries*, Washington, D.C.: National Academy Press, 1993, pp. 42–69.

Muhammed, Amir, "Population and Agricultural Resources in the Developing Countries," in Bernardo Colombo, Paul Demeny, and Max F. Perutz, eds., *Resources and Population: Natural, Institutional, and Demographic Dimensions of Development*, Oxford, U.K.: Clarendon Press, 1996, pp. 88–96.

Myers, Norman, "Environmental Refugees," *Population and Environment: A Journal of Interdisciplinary Studies*, Vol. 19, No. 2, 1997, pp. 167–182.

Oberthür, Sebastian, *Production and Consumption of Ozone-Depleting Substances, 1986–1995*, Bonn, Germany: Deutsche Gesellschaft für Technische Zusammenarbeit, 1997, p. 30.

O'Neill, Brian C., F. Landis MacKellar, and Wolfgang Lutz, *Population and Climate Change*, Laxenburg, Austria: International Institute for Applied Systems Analysis, 2000.

Orians, Carlyn E., and Marina Skumanich, *The Population-Environment Connection: What Does It Mean for Environmental Policy?* Battelle Seattle Research Center, prepared for Futures Studies Unit, Office of Policy Planning and Education, U.S. Environmental Protection Agency, December 1995.

Pebley, Anne R., "Demography and the Environment," *Demography*, Vol. 35, No. 4, 1998, pp. 377–389.

Perez, S., and F. Protti, *Comportamiento Del Sectore Forest a I Durante el Periodo 1950–1977*, San José, Costa Rica: Oficina de Planificacion Sectorial Agropecuaria, 1978.

Perrings, Charles, "Income, Consumption and Human Development: Environmental Linkages," in *Consumption for Human Development, Human Development Report, Background Papers,* New York: Human Development Report Office, United Nations Development Program, 1998, pp. 151–212.

Population Reference Bureau (PRB), "A World Profile of Environment and Population," in *Population Today,* Vol. 26, No. 9, September, 1998a.

_____, "UN Projections Assume Fertility Decline, Mortality Increase," *Population Today,* Vol. 26, No. 12, December 1998b.

_____, *World Population: More Than Just Numbers,* Washington, D.C.: PRB, 1999.

_____, *World Population Data Sheet: Demographic Data and Estimates for the Countries and the Regions of the World,* Washington, D.C.: PRB, 2000.

Postel, Sandra, *Pillar of Sand,* New York: W. W. Norton, 1999.

Qutub, Syed Ayub, "Rapid Population Growth and Urban Problems in Pakistan," *Ambio,* Vol. 21, No. 1, 1992, pp. 46–49.

Rahman, X., "Groundwater Contamination in Rapidly Growing Asian Cities," Stockholm Water Front, June 1, 1995, as presented in "Solutions for a Water-Short World," Johns Hopkins University School of Public Health, Population Information Program, Center for Communications Programs, Vol. 26, No. 1, 1995, pp. 8–9.

Rasmuson, Marianne, and Rolf Zetterstrom, "World Population, Environment and Energy Demands," *Ambio,* Vol. 21, No. 1, 1992, pp. 70–74.

Raven, Peter H., and Jeffrey A. McNeely, "Biological Extinction: Its Scope and Meaning for Us," in G. D. Lakshman and J. A. McNeely, eds., *Protection of Global Biodiversity: Converging Strategies,* Durham, N.C.: Duke University Press, 1998, pp. 13–32.

Repetto, Robert, "The 'Second India' Revisited: Population Growth, Poverty, and Environment over Two Decades," in R. K. Pachauri and L. F. Qureshy, eds., *Population, Environment, and Develop-*

ment, New Delhi: Tata Energy Research Institute, 1997, pp. 153–175.

Repetto, Robert, and Thomas Holmes, "The Role of Population in Resource Depletion in Developing Countries," *Population and Development Review,* Vol. 9, No. 4, 1983, pp. 609–632.

Rock, Michael T., "The Stork, the Plow, Rural Social Structure, and Tropic Deforestation in Poor Countries?" *Ecological Economics,* Vol. 18, No. 2, 1996, pp. 113–131.

Rosenzweig, Cynthia, and Martin L. Parry, "Potential Impacts on Climate Change on World Food Supply: A Summary of a Recent International Study," in Shridath Ramphal and Steven W. Sinding, eds., *Population Growth and Environmental Issues,* Westport, Conn.: Praeger Publishers, 1996.

Rosero-Bixby, Luis, and Alberto Palloni, "Population and Deforestation in Costa Rica," *Population and the Environment: A Journal of Interdisciplinary Studies,* Vol. 20, No. 2, 1998, pp. 149–185.

Rudel, Thomas K., "Population, Development, and Tropical Deforestation: A Cross-National Study," *Rural Sociology,* Vol. 54, No. 3, 1989, pp. 327–338.

Sader, S. A., and A. T. Joyce, "Deforestation Rates and Trends in Costa Rica, 1940 to 1983," *Biotropica,* Vol. 20, No. 1, 1988, pp. 11–19.

Shapiro, David, "Population Growth, Changing Agricultural Practices, and Environmental Degradation in Zaire," *Population and Environment: A Journal of Interdisciplinary Studies,* Vol. 16, No. 3, 1995, pp. 221–236.

Simon, Julian L., *The Ultimate Resource,* Princeton, N.J.: Princeton University Press, 1981.

_____, *The Ultimate Resource 2,* Princeton, N.J.: Princeton University Press, 1996.

Skole, D. L., W. H. Chomentowski, W. A. Salas, and A. D. Nobre, "Physical and Human Dimensions of Deforestation in Amazonia," *BioScience,* Vol. 44, No. 5, 1994, pp. 314–322.

Southwick, Charles H., *Global Ecology in Human Perspective,* New York: Oxford University Press, 1996.

Stern, Paul C., Thomas Dietz, Vernon W. Ruttan, Robert H. Socolow, and James L. Sweeney, eds., *Environmentally Significant Consumption: Research Directions,* Committee on the Human Dimensions of Global Change, Commission on Behavioral and Social Sciences and Education, National Research Council, Washington, D.C.: National Academy Press, 1997.

Stern, Paul C., Oran R. Young, and Daniel Druckman, eds., *Global Environmental Change: Understanding the Human Dimensions,* Committee on the Human Dimensions of Global Change, Commission on the Behavioral and Social Science and Education, National Research Council, Washington, D.C.: National Academy Press, 1992.

Subedi, Bhim P., "Population and Environment Interrelationships: The Case of Nepal," in R. K. Pachauri and L. F. Qureshy, eds., *Population, Environment, and Development,* New Delhi: Tata Energy Research Institute, pp. 191–214.

Switzer, J. V., and G. Bryner, *Environmental Politics: Domestic and Global Dimensions,* New York: St. Martin's Press, 1996.

Tett, Simon, John F. B. Mitchell, David E. Parker, and Myles R. Allen, "Human Influence on the Atmospheric Vertical Temperature Structure: Detection and Observations," *Science,* Vol. 274, November 15, 1996, pp. 1170–1173.

Thomas, N., "Review of UNFPA, Population, Resources, and the Environment: The Critical Challenges," *Population Studies,* Vol. 46, No. 3, pp. 559–560.

Torikai, Yukihiro, "Development of the Philippine Frontier—Labor Absorption and Internal Migration to Palawan Province," *Southeast Asian Studies,* Vol. 31, No. 3, December 1993, pp. 255–284.

United Nations, *Agenda 21: Programme of Action for Sustainable Development, Rio Declaration on Environment and Development,* New York: United Nations, 1992.

_____, *Population, Environment and Development,* Proceedings of the United Nations Expert Group Meeting on Population, Environment and Development, United Nations Headquarters, January 20–24, 1992, convened as part of the substantive preparations for the International Conference on Population and Development, 1994.

_____, *Urban Agglomerations 1996,* wall poster, New York, 1997a.

_____, *Urban and Rural Areas 1996,* wall poster, New York, 1997b.

_____, *Briefing Packet: World Population Estimates and Projections, 1998 Revision,* New York, 1998a.

_____, "World Population Projections to 2150," New York, February 1, 1998b.

_____, *World Urbanization Prospects: The 1996 Revision,* New York, 1998c.

_____, "World Population Would Stabilize at Nearly 11 Billion by Year 2200," press release, Population/656, New York, February 2, 1998d.

_____, *The Demographic Impact of HIV/AIDS,* report on the Technical Meeting, New York, November 10, 1998, Population Division, Department of Economic and Social Affairs, ESA/P/WP.152, February 1999.

United Nations Development Programme (UNDP), *Consumption for Human Development: Background Papers,* Human Development Report, New York, 1998.

United Nations Environment Programme (UNEP), *Global Environmental Outlook,* New York: Oxford University Press, 1997.

United Nations Population Fund (UNFPA), *Population, Resources and The Environment,* London: UNFPA, 1991.

_____, *The State of World Population, 1998,* URL: http://www.unfpa.org/, 1999a (retrieved March, 1999).

_____, *The State of World Population, 1999,* URL: http://www.unfpa.org/, 1999b (retrieved March, 1999).

U.S. Department of Agriculture (USDA), *National Resources Inventory, 1997, Summary Report,* Natural Resources Conservation Service, Iowa State University Statistical Laboratory, Washington, D.C.: USDA, December 1999a, URL: http://www.nhq.nrcs.usda.gov/CCS/NRIrlse.html (retrieved December 20, 1999).

_____, *Preserving the Health of the Land: America's Conservation Challenge,* Natural Resources Conservation Service, Iowa State University Statistical Laboratory, Washington, D.C.: USDA, December 1999b, URL: http://www.nhq.nrcs.usda.gov/NRI/1997/other_reports/4-Pager.pdf (retrieved December 1999).

Wernick, Iddo K., "Consuming Materials: The American Way," in Paul C. Stern, Thomas Dietz, Vernon W. Ruttan, Robert H. Socolow, and James L. Sweeney, eds., *Environmentally Significant Consumption,* Committee on the Human Dimensions of Global Change, Commission on Behavioral and Social Sciences and Education, National Research Council, Washington, D.C.: National Academy Press, 1997, pp. 29–39.

Weyant, John, and Yukio Yanigisawa, "Energy and Industry," in Steve Rayner and Elizabeth L. Malone, eds., *Human Choice & Climate Change,* Vol. 2, Columbus, Ohio: Battelle Press, 1998, pp. 203–289.

Whitmore, Thomas M., B. L. Turner II, Douglas L. Johnson, Robert W. Kates, and Thomas R. Gottschang, "Long-Term Population Change," in B. L. Turner II, William C. Clark, Robert W. Kates, John F. Richards, Jessica T. Mathews, and William B. Meyer, eds., *The Earth as Transformed by Human Action: Global and Regional Changes in the Biosphere over the Past 300 Years,* Cambridge, Mass.: Cambridge University Press, 1990, pp. 25–40.

Wilson, E. O., *The Diversity of Life,* Cambridge, Mass.: The Belknap Press of Harvard University Press, 1992.

Wolman, M. Gordon, "Population, Land Use, and Environment: A Long History," in *Population and Land Use in Developing Countries,* National Research Council, Washington, D.C.: National Academy Press, 1993.

Wood, Charles H., and Stephen G. Perz, "Population and Land-Use Changes in the Brazilian Amazon," in Shridath Ramphal and

Steven W. Sinding, eds., *Population Growth and Environmental Issues*, Westport, Conn.: Praeger Publishers, 1996.

World Bank, *World Development Report 1992: Development and the Environment*, New York: Oxford University Press, 1992.

_____, *World Development Report 1999/2000. Entering the 21st Century: The Changing Development Landscape*, New York: Oxford University Press, 1999, URL: http://www.worldbank.org/wdr/2000/fullreport.html (retrieved December 1999).

World Commission on Environment and Development (WCED), *Our Common Future*, Oxford, U.K.: Oxford University Press, 1988.

World Conservation Monitoring Centre (WCMC), *Global Biodiversity: Status of the Earth's Living Resources*, London: Chapman and Hall, 1992, pp. 196–197.

World Health Organization, *The World Health Report 1998*, Geneva, Switzerland: WHO, 1998a.

_____, "3 Billion People Worldwide Lack Sanitation Facilities: WHO Strategy on Sanitation for High-Risk Communities," January 29, 1998, Press Release WHO/18, 1998b.

_____, "Point of Fact #77," Division of Diarrheal and Acute Respiratory Disease Control, 1998c.

World Health Organization and UNEP, *Urban Air Pollution in Megacities of the World*, Oxford, U.K.: Blackwell Reference, 1992, p. 39.

World Meteorological Organization (WMO), *Comprehensive Assessment of the Freshwater Resources of the World*, Geneva, Switzerland: WMO, 1997.

World Resources Institute, *World Resources: A Guide to the Global Environment, 1998–99*, The World Resources Institute, UNEP, United Nations Development Programme, The World Bank, New York: Oxford University Press, 1998.

ADDITIONAL REFERENCES

GENERAL

Commoner, Barry, "Rapid Population Growth and Environmental Stress," *International Journal of Health Services*, Vol. 21, No. 2, 1991, pp. 199–227.

Demeny, Paul, "Tradeoffs Between Human Numbers and Material Standards of Living," in K. Davis and M. Bernstam, eds., *Resources, Environment, and Population: Present Knowledge and Future Options*, New York: Oxford University Press, 1991, pp. 408–421.

Hogan, Daniel Joseph, "The Impact of Population Growth on the Physical Environment," *European Journal of Population*, Vol. 8, 1992, pp. 109–123.

Keyfitz, Nathan, "Population Growth, Development and the Environment," *Population Studies*, Vol. 50, 1996, pp. 335–359.

Livernash, Robert, and Eric Rodenburg, *Population Change, Resources, and the Environment*, Population Bulletin, New York: Population Reference Bureau, 1998.

McNicoll, Geoffrey, "On Population Growth and Revisionism: Further Questions," *Population and Development Review*, Vol. 21, 1995, pp. 307–340.

Myers, Norman, *Population, Resources and the Environment: The Critical Challenges*, United Nations Population Fund, London: Banson Publishing, 1991.

National Research Council, *Global Environmental Change: Understanding the Human Dimensions,* Paul C. Stern, Orban R. Young, and Daniel Druckman, eds., Committee on the Human Dimensions of Global Change, Commission on the Behavioral and Social Sciences and Education, Washington, D.C.: National Academy Press, 1992.

_____ *Global Environmental Change: Research Pathways for the Next Decade,* Committee on the Human Dimensions of Global Change, Commission on the Behavioral and Social Sciences and Education, Washington, D.C.: National Academy Press, 1999.

Ness, Gayl D., William D. Drake, and Steven R. Brechin, eds., *Population-Environment Dynamics: Ideas and Observations,* Ann Arbor, Mich.: University of Michigan Press, 1993.

Panayotou, Theodore, "An Inquiry into Population, Resources, and the Environment," in D. Ahlburg, A. C. Kelley, and K. O. Mason, eds., *The Impact of Population Growth on Well-Being in Developing Countries,* New York: Springer-Verlag, 1996, pp. 259–298.

Preston, Samuel H., *Population and the Environment: From Rio to Cairo,* International Conference on Population and Development, International Union for the Scientific Study of Population, distinguished lecture series on population and development, 1994.

Preston, Samuel H., "The Effect of Population Growth on Environmental Quality," *Population Research and Policy Review,* Vol. 15, April 1996, pp. 95–108.

Repetto, Robert, "Renewable Resources and Population Growth: Past Experiences and Future Prospects," *Population and Environment: A Journal of Interdisciplinary Studies,* Vol. 10, No. 4, 1989, pp. 221–236.

Science, issue devoted to Human-Dominated Ecosystems, Vol. 277, July 25, 1997.

RESEARCH METHODS

Liverman, Diana, Emilio F. Moran, Ronald R. Rindfuss, and Paul C. Stern, eds., *People and Pixels: Linking Remote Sensing and Social*

Science, Committee on the Human Dimensions of Global Change, Commission on Behavioral and Social Sciences and Education, National Research Council, Washington, D.C.: National Academy Press, 1998.

ENVIRONMENT INFLUENCES ON DEMOGRAPHIC PROCESSES

Mortality. The links among environmental factors, health, and mortality are probably the best-studied among the environmental influences on demographic processes, with the health impacts of air and water pollution having received the most research attention.

Lipfert, Frederick W., "Air Pollution and Human Health: Perspectives for the '90s and Beyond," *Risk Analysis,* Vol. 13, No. 2, 1997, pp. 137–146.

Nash, Linda, "Water Quality and Health," in P. H. Gleick, ed., *Water in Crisis: A Guide to the World's Fresh Water Resources,* New York: Oxford University Press, 1993.

Fertility. Only a handful of studies have attempted to make the link between environmental conditions and fertility, and they remain fairly speculative.

Kalipeni, Ezekiel, "Demographic Response to Environmental Pressure in Malawi," *Population and Environment: A Journal of Interdisciplinary Studies,* Vol. 17, No. 4, 1996, pp. 285–308.

Takahashi, Shinichi, "The Demographic Transition in Rural Areas of Northeastern Thailand: Two Regimes of Population," *Journal of Population Studies,* No. 20, May 1997, pp. 49–63.

Migration. Environmental factors are among the many "push" and "pull" forces acting in the migration decisionmaking process.

Hunter, Lori M., "The Association Between Environmental Risk and Internal Migration Flows," *Population and Environment: A Journal of Interdisciplinary Studies,* Vol. 19, No. 3, pp. 247–277.

McGranahan, David A., *Natural Amenities Drive Rural Population Change,* Food and Rural Economics Division, Economic Research

Service, U.S. Department of Agriculture, Agricultural Economic Report No. 781, 1999.

Myers, Norman, "Environmental Refugees," *Population and Environment: A Journal of Interdisciplinary Studies*, Vol. 19, No. 2, 1997, pp. 167–182.

Thiam, Babaly, "Environmental Impact on Migration and on the Spatial Redistribution of the Population," in *Population, Environment and Development*, Proceedings of the United Nations Expert Group Meeting on Population, Environment and Development, United Nations Headquarters, January 20–24,1992, convened as part of the substantive preparations for the International Conference on Population and Development, Chapter 15, 1994, pp. 175–185.